The Long-Term Fate of Invasive Species

Arne Jernelöv

The Long-Term Fate of Invasive Species

Aliens Forever or Integrated Immigrants with Time?

Springer

Arne Jernelöv
Swedish Institute for Future Studies
Jarpas, Sweden

ISBN 978-3-319-55395-5 ISBN 978-3-319-55396-2 (eBook)
DOI 10.1007/978-3-319-55396-2

Library of Congress Control Number: 2017937316

Printed on acid-free paper

This Springer imprint is published by Springer Nature
The registered company is Springer International Publishing AG
The registered company address is: Gewerbestrasse 11, 6330 Cham, Switzerland

Preface

Alien, in the context of invasive species, means that the species, subspecies, or group with some specific genetic traits has (recently) been introduced by humans to a region, usually a continent or island, to which it didn't spontaneously spread. Mostly, alien is seen as a prerequisite for invasive.

The term invasive, when it is applied to a species other than human, is used with several different meanings. In the broadest understanding of the term, it means a species that, with the help of humans, establishes itself in an area outside its native range. In a somewhat more narrow understanding, it should not only survive for generations in the new territory but also undergo a drastic population expansion. A further limitation of the term comes when a damage criterion is introduced, mostly expressed in economic terms. Thus, the most widely used, human-centered definition reads "An invasive species is a plant or animal that is not native to a specific location; and has a tendency to spread, which is believed to cause damage to the environment, human economy and/or human health." In practice, as national governments and authorities are key players, the word "native" is understood to mean "from our country." Thus, the house finch, native to California and brought by humans to the US east coast, from where it has spread all over the USA, Mexico, and southern Canada and now numbers over a billion, is not classified as alien or invasive in the USA and Mexico. However, the house sparrow, with much lower numbers, but coming from Europe, is classified as invasive. Whether the house finch shall be seen as an invasive alien in Canada becomes a different administrative and philosophic question, since the original west coast population lately has spread north on its own and entered British Columbia, while the human-moved eastern population has spread to Ontario, Quebec, and the Atlantic provinces. Another example of a species that has been moved by humans within a country and then spread itself from the new location to other countries is the raccoon dog (*Nyctereutes procyonoides*). It was brought by humans from eastern Siberia to European Russia and Ukraine during Soviet times and then spread naturally from there to the Baltic states, Finland, and Sweden. It is regarded as alien and invasive in the latter countries, but not in European Russia.

Sometimes, a moral argument comes up for forgiving the invasiveness of a species when it prospers in an area to which humans brought it—when humans altered its native area to the point where the species can no longer survive there. The snowdrop (*Galanthus nivalis*) is about to become such a species thriving well on the British Isles and in Scandinavia, to where humans brought it, but fighting for survival in its native Continental Europe. Global warming is likely to produce quite a few such species. Are they to be seen as aliens that should be exterminated if possible, or are they refugees from our manipulations of the environment that we are morally obliged to give a new home in the world we altered?

In the definition of invasive cited above, the expression "is believed to" is used instead of the much stricter "has been showed to" or even more so "has been proven to." This can possibly be seen as an off-shoot of the wordings used in the definition of the precautionary principle from the 1992 Rio Summit, where it is said that "Where there are threats of serious or irreversible damage, lack of full scientific certainty shall not be used as a reason to postponing cost-effective measures to prevent environmental degradation."

There is, however, a clear further weakening of the requirement to demonstrate damage, if "lack of full scientific certainty" is seen as the mother of "is believed to" and the question of who is authorized to be the believer immediately comes to mind.

Another aspect, if the precautionary principle is invoked, is the wording "cost-effective measures to prevent environmental degradation." This would seem to imply that measures to reduce or eradicate invasive species should only be undertaken if the cost of those measures is lower than the damage caused by the invader. (The methods used to calculate economic and ecological damage will be commented on in the concluding chapter.)

In this book, the term invasive is used without any damage criteria. I have used the term for species, which conquered new territories after having been introduced by humans, and that increased dramatically in numbers there. There are many of those, so to arrive at the 18 cases presented here, further selection criteria have been applied. First and foremost among those is time since introduction, as the long-term fate is the theme of the book. The order of the chapters with case studies also reflects this as they are presented following the year of introduction with the oldest, the earthworms, first.

The idea has also been to select examples from different groups of animals and plants, and organisms coming from and invading different geographical areas. The result comprises 18 cases: two parasitic fungi, three plants, six invertebrates (out of which three are insects), one amphibian, two birds, and four mammalians. Invaded areas are Africa, Asia, Australia, Europe, North America, Pacific Islands, and South America. The invaders come from Africa, Asia, Europe, North America, and South America. The invaded areas include terrestrial systems, as well as fresh and marine waters, in temperate and tropical areas in the northern and southern hemispheres.

Jarpas, Sweden Arne Jernelöv

Acknowledgments

The chapters in this book contain some seven hundred references, about 15% of the total number of publications I reviewed in the writing process. Most of them come from easily available scientific publications, but quite a few were found in much more difficult to access places such as 150-year-old articles in local small-towns, long-since-closed newspapers, in-house publications at various laboratories and institutes (so-called gray literature), articles in less common languages, or social network media postings. To retrieve it all has been a most challenging librarian task, and without the unwavering support of Michaela Rossini and her colleague Natalia Ovtchinnikova at the library of the International Institute for Applied Systems Analysis (IIASA), it would not have been possible for me to write this book.

I'm also indebted to Muki Jernelöv, my wife, for putting a lighter, native-speaker touch to the language and for editing the manuscript.

In addition, Bo Söderström, editor of *Ambio*, the environmental journal of the Royal Swedish Academy of Sciences, provided early inspiration for the book through our discussions about the fate of sparrows and finks in North America.

Contents

Earthworms in North America

After the latest ice ages, earthworms were, by and large, absent from North America. Only in some sheltered regions that did not experience deep permafrost did the native species survive. It is a little known and seldom acknowledged and considered fact, however, that these earthworms didn't spread far after the glaciations. Thus, the prevailing earthworms found practically everywhere in North America and generally seen as natural, desirable, and self-evident components of the terrestrial fauna were more or less unintentionally brought in by European immigrants to become very successful invaders.

The Earthworms in Their Native Habitats

Earthworms are tube-shaped, segmented organisms that live in soil and feed on organic matter. They are invertebrates and thus lack a skeleton, but possess liquid-filled chambers which act as a hydraulic skeleton that can provide stability to whichever part of the body needs it at a given moment. They are very flexible, with longitudinal as well as transversal muscles. The digestive channel goes through the full length of the animal, and they are hermaphrodites, so each individual has both female and male sex organs. When mating, they reciprocally fertilize each other. Their coloration is mostly in the range of red-violet to gray-blue. There are globally some 2000 species of earthworms. In Europe, one of the most important groups is Lumbricidae, to which the common earthworm, *Lumbricus terrestris*, belongs. It can reach a length of 30 cm, which isn't much compared to the largest species, the 7 m long South African gigantic earthworm.

The workings of earthworms in the terrestrial ecosystem are well known. They live in the boundary layer between plant material such as decomposing leaves and grass on the surface and the mineral soil beneath. They live off of organic matter and act to speed up its decomposition by mixing more persistent parts of it such as humus with the inorganic soil components deeper down. In the process, the tunnels

A. Jernelöv, *The Long-Term Fate of Invasive Species*,
DOI 10.1007/978-3-319-55396-2_1

they dig facilitate aeration and drainage of the soil, which in turn further enhance microbial activity and release of nutrients.

Earthworms are generally well regarded in Europe and Asia, among farmers as well as scientists. Charles Darwin [1], who published a book about them in 1881, put great emphasis on their importance. “It may be doubted,” he wrote, “if there are any other animals which have played such an important part in the history of the world as these lowly organized creatures.”

One hundred and thirty years later, BBC Nature sang their praise: “Earthworms are the world’s unsung heroes. They loosen and mix up the soil, break down and recycle decaying plant matter and fertilize the soil by bringing nutrients closer to the surface. Birds often rely upon worms as a primary source of food.”

The common earthworm *Lumbricus terrestris*

As education for children, Skansen, an open-air museum and zoo in Stockholm, Sweden, writes on its home page: “Earthworms are small but very good farmers. They mix the soil and drag dead leaves down into the ground. Their tunnels go deep down and through them air and water can reach the roots of plants. They benefit other animals and plants.”

Introduction and Spread

Although hardly any records exist, it is obvious that earthworms of many different species were introduced to North America on many different occasions and to many different places. It is also very easy to imagine how it happened. Settlers from Europe often brought all sorts of plants with them from their home countries and with them soil around the roots. In the soil, there would have been earthworms.

However, one person is first and foremost seen as the one who happened to introduce earthworms to the New World. That person was John Rolfe. His place in the history of European settlement of America is secure also for several other reasons, even if it could be argued that his introduction of the earthworm was the action with the most far-reaching and long-term consequences.

John Rolfe was a settler in Jamestown, Virginia, the first successful English colony in North America, and he was the person that more than any other made it so. He married Pocahontas, the daughter of the local Indian chief, Powhatan, thereby securing peace and food assistance that were vital for the survival of the settlers. He also introduced and started the cultivation of the tobacco plant *Nicotiana tabacum*. When he returned to England in 1616, he brought with him not only his Indian wife but also a large cargo of tobacco from his farm at Jamestown. Both were a huge success in London. John Rolfe and Jamestown became rich from the regular shipments of tobacco to England, and the small colony became a boomtown with a rapidly growing population. The ships that came to fetch the "brown gold" carried ballast of stone, gravel, and dirt, which was dumped onto the beach to make room for the bales of dried tobacco leaves. With the ballast came the earthworms—or so the story goes.

It is well documented that the settlers in Jamestown during the first decade found it hard to grow any of the plants they brought with them. Historians have attempted to explain this and mostly looked for socioeconomic explanations, one of them being that few of the settlers had been farmers back in England. That might well have been a factor, as was in all likelihood the severe drought of the summer of 1609, but these circumstances may not be the whole explanation. Whatever the reasons, the consequences were dramatic. During the "winter of starvation" 1609–1610, 440 out of the original 600 settlers died. Survivors were abandoning the colony, when a ship carrying new immigrants, among them John Rolfe, arrived with provisions and seeds. But also in subsequent years, harvests were minimal, not only of grain but also of cabbage, turnips, and other vegetables. The only crop the settlers planted that grew well from the start was tobacco. They could trade and get maize from the Indians, but for the settlers, it seems they didn't really attempt to grow it for themselves. Eventually, little by little, their vegetable gardens finally started to produce.

Let's return to the reasons for the settlers' failure to produce adequate food. If the low crop yield in 1609 was really due to drought, why did they continue to be unsuccessful in the following years that enjoyed normal precipitation? If it was lack of farming skills, why did they immediately succeed with tobacco? One possible explanation is that tobacco and the maize that the Indians grew are American plants, the domestication of which had taken place in the absence of earthworms. The crops the Jamestown's settlers planted that failed to thrive were introduced from Europe and Asia, where earthworms have long had a decisive influence over the structure and quality of soils.

Back in Jamestown, earthworms slowly spread and started to give the soils in and around the colony a more European-like structure, which facilitated Eurasian plants to establish themselves there. This was, of course, the case not only for agricultural plants.

In a pattern resembling that of the highly suspect ballast in Jamestown, other, later, settlers came with plants and soil containing earthworms from their home countries. These became established and spread, in the slow way of worms, from their respective places of release. As a rule of thumb, earthworms on their own can extend their territory with a radius of maybe 10 m a year; so for their further spread inside North America, humans also played a key role. Besides gardeners, many of whom wanted earthworms to improve their soil and actively sought them out much of the spread was unintentional via transported plants and soil. Fishermen also played an important role, as can be concluded from the prevalence of earthworms around angling lakes and rivers. Leftover worms were simply released at the fishing sites. This resulted in that earthworms are now present in most North American habitats suitable for them. The different species, however, have a very uneven distribution, which has more to do with their haphazard pattern of introduction than their individual competiveness vis-à-vis each other in a given environment. One can conclude that as a group, earthworms have more or less conquered the continent, but the individual species have not yet reached their final distribution and, often enough, not yet met each other as competitors.

Notwithstanding the foregoing, to say that North America completely lacked earthworms after the latest glaciations is not quite correct. A not-so-small number of species, including a couple in the family Lumbricidae, did survive, but with a fairly limited geographic distribution outside the ice-free zones where they had hung on. With regard to the number of species of earthworms, the newcomers make up about a third, but with regard to number of individuals or biomass, they totally dominate.

Ecological Effects

The ground in a forest or meadow without earthworms is characterized by a layer of leaves and/or dead grass, often several decimeters and occasionally up to half a meter thick, representing several years of litterfall. Below this, there is mostly a thin

The ground in a forest where maple trees dominate and earthworms are absent. The leaf layer has been flattened after several months of snow cover. The leaves stick together, and the ones on the surface have been bleached, which indicates that no worms crawl through and disrupt it (Photo: G. Schlaghamensky and Kalev Jogiste [2])

A 20 cm long soil profile with a thick leaf layer to the right, fragmented leaves (the brown layer with light spots), followed by a layer of humus (black), and the beige mineral layer. The soil sampler is placed on the ground in question (Photo: G. Schlaghamensky)

layer of nutrient-rich compost soil and, underneath that, a clay or other mineral soil poor in nutrients [2].

Where earthworms are present, the previous year's litter generally disappears before autumn brings down another load. Under the thin layer of not-yet-degraded

The earthworms (*Lumbricus terrestris*, *L. rubellus*, and *Aporrectodea* sp.) have arrived and started to change the soil profile. A thin layer of leaves on the surface covers an even thinner, almost invisible, strata of leaf fragments below which the black humus has started to mix with the beige mineral soil (Photo: G. Schlaghamensky)

Here, the earthworms have changed the soil profile completely. The top layer no longer consists of leaves, as the worms quickly consume them, but of leaf fragments and twigs. Under this the soil layers are totally mixed, as they had been plowed (Photo: G. Schlaghamensky)

organic matter, there is a mixing zone of humus and mineral soil, on top of the clay or sand that lies beneath.

Earthworms transformed the soil of North American forests and grasslands with significant consequences for the viability and competiveness of many plants. With a thick layer of leaves and grass on the ground, budding plants must send out long roots to reach the moist and nutrient-rich humus, which, though rich in nutrients, it is only a thin layer. In the earth below, the concentrations of essential nutrients such as phosphorus and nitrogen are low. To succeed, the plants thus need to concentrate their nutrient-sucking root treads in this soil layer. For bigger plants such as trees, however, roots also have another essential function, which is to anchor and stabilize. To achieve this in a thin soil layer near the surface, the root system must be very dispersed and far-flung. A typical case is the redwood tree (*Sequoia sempervirens*). Deep roots would anchor the tree but be of little use for nutrient and water uptake. Native North American plants are generally adapted to this type of soil that used to dominate the continent.

In the thick, multiyear layer of leaves and grass lives a rich fauna of organisms. Insects such as millipedes, wood lice, springtails, mites, and spiders often dominate, but lizards, snakes, frogs, salamanders, mice, and shrews are also abundant. When the special environment that provides them with both food and protection shrinks in both time and area, the populations of these species often decline drastically [3–6].

Often, the earthworm densities are highest in the first years after invasion, when a large accumulated food supply is at hand. Later, densities adjust to type and quantity of recurring litterfall.

In cases where native North American earthworms co-occur with alien species and likely face some degree of competition, some reports suggest that the invader together with land-use change and habitat fragmentation had a strong negative impact on the native species [7], while many others found no such detrimental interaction between native and alien earthworms [4, 8].

Interacting Alien Species

While the Eurasian plants are evolutionary adjusted to the types of soils that earthworms create and maintain, their North American counterparts are not. When the structure of soil is altered, the competiveness of native species decreases, and the spread of vegetation from other continents with similar climate is facilitated [9].

That earthworms facilitate the establishment of Eurasian plant species by altering the structure of the soils can be seen as an established fact. To what extent this also means that they are important accomplices for the invasive success of, e.g., Japanese knotweed (*Fallopia japonica*), cheat grass (*Bromus tectorum*), and kudzu (*Pueraria montana*) in the USA, is less clear. In some cases, invading plants seem to return the favors and promote earthworms in return. For instance, biomass and abundance of invasive earthworms increased in plots dominated by European buckthorn (*Rhamnus cathartica*) probably as the low C/N ratio in their leaf liver is favorable for the worms (Heneghan and Steffen, unpublished) [5].

This general argument about invasive species facilitating for compatriots can be extended to the competitiveness of non-plant species as well. For example, introduced European birds such as starlings have earthworms high on their list of preferred food items and have evolved skills to locate and catch them, which the North American species had no general use for until the worms appeared.

Although this general reasoning makes sense (at least to me), it should be pointed out that I've found no specific scientific studies supporting or refuting the hypothesis of positive interaction of earthworms with other alien species than plants.

A study of the European earthworm *Lumbricus rubellus* and the Asian *Amynthas agrestis* in Tennessean soil found that a higher degree of food flexibility gave the Asian invader the upper hand in the competition with the European one [10]. On the other hand, a study of competition between the European earthworm, *Lumbricus rubellus*, and the Asian *Amynthas hilgendorfi* found no negative effect of the presence of one species on the other [11].

Economic and Human Health Impacts

An overwhelming part of all articles dealing with earthworms in North America describes them in similar positive terms as corresponding articles in Europe and Asia do there. The number of earthworms per surface area is often seen as a direct measure of agricultural soil fertility [12–15] and promoters of "organic farming" almost always stress that with their methods of cultivation, the number of earthworms will increase significantly. These in turn will not only help fertilize the soil, they argue, but also increase soil porosity so that more water can be stored, decreasing runoff and the risk of floods [16].

A very extensive review of the literature on the role of earthworms in soil formation and provision of ecosystem services, with 252 references [17] gave in the end a very positive picture of the contribution of these "lowly creatures," as Darwin termed them. However, in several studies included in the review, the positive picture was less clear, and in a few cases, the opposite conclusion was actually drawn.

Several attempts have been made to calculate the value of the ecosystem services provided by earthworms in economic terms. An early and often cited study [18] arrived at a global figure of US$75 billion a year for soil improvements. If one approximates this figure in relation to North America's share of global land under cultivation, about 15%, one comes to an annual value of US$11.25 billion.

Another approach to the question was used by the professors Steve Wratten and Ross Cullen at Lincoln University, New Zealand [19]. Their calculations were based on the value of fertile topsoil and the amount of such that earthworms produce in a year. The result would then be proportionate to the density of worms in the soil. Their standard example was 1.3 tons of earthworms per ha, which gave a value of US$54 per year. Recalculated to North American prices and size of land under agriculture (4.7 million km^2) and using the same earthworm density, the figure would be $12 billion annually.

This, however, is far from the whole picture. During the last decade, forest scientists in particular have started to describe earthworms in negative terms [20]. One part of their argument has been, contrary to many other studies some of which are referred to above, that the accelerated degradation of organic matter caused by the earthworms leads to an increased loss of nutrients. Their most important argument, however, is that the increased difficulties for native plants, foremost trees, to root themselves in the soil that earthworms create—a claim about which there is full consensus—have not only ecological but also economic repercussions [21].

As this article also highlights, it is not only European earthworms that have invaded North America but also Asian ones, in particular the *Amynthas* species, which arrived late in the nineteenth century. Some vocal scientists see them as an even worse threat than their European counterparts to the remaining natural forests in North America.

In summary, though, the positive economic effects on North American agriculture and horticulture are so strong that it's highly unlikely that the effects on forestry, when they in the future get assessed, can change the picture of an invasion with a clear black bottom line.

There are some articles that speculate about the possibility that earthworms may carry certain bacteria and viruses that could be detrimental to human health and thereby spread diseases, but there are no actual case studies to substantiate the claim, besides reports of diarrhea after consumption of mouthfuls of earthworms.

Control Efforts

There have been no reported serious attempts to eradicate earthworms once they got established. There's some information given to anglers in particular, arguing that they should not dump leftover bait worms in still earthworm-free areas. Generally, though, much more efforts are spend actively spreading them than preventing their spread.

Current Status and Likely Future

Although earthworms today are spread over most of North America except for the far north, mountain areas, and deserts, where they can't live, and some forest areas they have not yet reached, the different species still mostly occur in a distribution pattern reflecting the haphazard nature of hundred thousand introductions to new sites.

With time, counted in centuries, different species, invasive and native, will coexist in the same areas and subdivide it into ecological niches. Their collective impact on soil will be even stronger, when the different species do what they are best at. In this context, it's worth pointing out that the same human-caused distribution factors that spread the invasive species are at work on the native ones and that they also transform the soil, when they enter an earlier earthworm-free area. The domination of the newcomers is however likely to continue.

Public Perception and Current Policies

The public perception of earthworms in North America is overwhelmingly positive. They are seen as indicative of fertile and healthy soil. Forest ecologists' recent concern has not (yet) had much impact either on the public at large or on decision makers. Thus, there are no specific anti-earthworm policies in place or under preparation [22].

References

1. Darwin, C. (1881). *The formation of vegetable mould through the action of worms, with observations on their habits*. London: John Murray.
2. Loss, S. R., Hueffmeier, R. M., Hale, C. M., Host, G. E., Sjerven, G., & Frelich, L. E. (2013). Earthworm invasion in northern hardwood forests: A rapid assessment method. *Natural Areas Journal, 33*(1), 21–30.
3. Eisenhauer, N., Partsch, S., Parkinson, D., & Scheu, S. (2007). Invasion of a deciduous forest by earthworms: Changes in soil chemistry, microflora, microarthropods, and vegetation. *Soil Biology and Biochemistry, 39*, 1099–1110.
4. Addison, J. A. (2009). Distribution and impacts of invasive earthworms in Canadian forest ecosystems. *Biological Invasions, 11*(1), 59–79.
5. Freilich, L. E., Hale, C. M., Scheu, S., Holdsworth, A. R., Heneghan, L., Bohlen, P. J., & Reich, P. B. (2006). Earthworm invasion into previously earthworm-free temperate and boreal forests. *Biological Invasions, 8*, 1235–1245.
6. Hendrix, P. F., Callahan Jr., M. A., & Drake, J. M. (2008). Pandora's box contained bait: The global problem of introduced earthworms. *Annual Review of Ecology, Evolution, and Systematics, 39*, 593–613.
7. León, Y. S.-d., & Johnson-Maynard, J. (2008). Dominance of an invasive earthworm in native and non-native grassland ecosystems. *Biological Invasions*. doi:10.1007/s10530-008-9347-6.
8. Hendrix, P. F., Baker, G. H., & Callahan Jr., M. A. (2006). Invasion of exotic earthworms into ecosystems inhabited by native earthworms. *Biological Invasions, 8*, 1287–1300.

9. Heneghan, L., Steffen, J., & Fagen, K. (2007). Interactions of an introduced shrub and introduced earthworms in an Illinois urban woodland: Impact on leaf litter decomposition. *Pedobiologia, 50*, 543–551.
10. Zhong, W., Hendrix, P. F., Snyder, B. A., Molina, M., Li, J., Rau, X., Siemann, E., & Shanglei, F. (2010). Dietary flexibility aids Asian earthworm invasion in North American forests. *Ecology, 91*(7), 2070–2079.
11. Greiner, H. G., Kashian, D. R., & Tiegs, S. D. (2012). Impacts of invasive Asian (*Amynthas hilgendorfi*) and European (*Lumbricus rubellus*) earthworms in a North American temperate deciduous forest. *Biological Invasions, 14*(10), 2017–2027.
12. Kladivko, E. J. *Earthworms and crop management* (AY-279 Soils/Management). West Lafayette, IN: Purdue University, Cooperative Extension Service.
13. Ramsay, J. A., & Hill, S. (1978). Earth worms: The agriculturalist's friends. *Macdonald Journal 39*(10), 6–8. Retrieved from www.eap.mcgill.ca/publications/eap6.htm
14. Earthworms Science Page, Garden Mosaics. Cornell University. Retrieved from www.garden-mosaics.cornell.edu/pgs/science/english/earthworm.htm
15. Invertebrate Identification Guide. University of Michigan. Retrieved from http://www.biokids.umich.edu/files/20491/InvertIDGuide2004.pdf
16. Edwards, C. A. *Earthworm ecology* (2nd ed.). Boca Raton, FL: CRC Press.
17. Blouin, M., Hodson, M., Aranda Delgado, E., Baker, G., Brussard, L., Butt, K. R., Dai, J., Peres, G., Yondoh, J. E., Cluzeau, D., & Butt, J. J. (2013). A review on earthworm impact on soil function and ecosystem services. *European Journal of Soil Science, 64*(2), 161–182.
18. Costanza, R., d'Arge, R., de Groot, R., Farber, S., Grasso, M., Hannon, B., Limburg, K., et al. (1997). The value of the world's ecosystem services and natural capital. *Nature, 387*, 253–260.
19. Wratten, S., Sandhu, H., Cullen, R., & Constanza, R. (2013). *Ecosystem services in agricultural and urban landscapes* (p. 218). QLD, Australia: Wiley-Blackwell.
20. Conover A. (2000, August). *Smithsonian Magazine*.
21. Coles, J., & Walker, M. (2012, September 24). Earthworm invasion: Aliens causing more harm than good? *BBC Nature*.
22. Callaham Jr., M. A., Gonzales, G., Hale, C. M., Henegham, L., Lachnicht, S. L., & Xiaoming, Z. (2006). Policy and management responses to earthworm invasions in North America. *Biological Invasions, 8*, 1317–1329.

Zebra Mussels in Western Europe and North America

Cautionary Tale

In a cautionary tale from the nineteenth-century Holland, a poor lockkeeper was faced with additional hardship when masses of mussels suddenly colonized the locks and prevented their operation. Removal of the mussels required hard work and was very time-consuming, and he suffered numerous, easily infected cuts. The poor man had to hire a helper, whom he could barely afford to feed. His luck changed, though, when he discovered that the chickens on his tiny farm, adjoining the locks, eagerly ate the small mollusks and that he, with an unlimited source of free food for them, could thus raise them in large numbers. In the end, he became a wealthy egg producer.

The story is told as one of a blessing in disguise, but there is far from being a consensus that zebra mussels qualify as such.

The Mussel in Its Native Habitat

The **zebra mussel** (*Dreissena polymorpha*) is a small freshwater **mussel** originating in lakes and rivers in Southern Russia. The name zebra mussel derives from the stripes on its triangular shells, although this color pattern is not always prevalent. They are filter feeders and attach themselves to hard substrates with strong byssus threads emanating from their umbo on the hinged side. They often form large colonies with mussels attached to one other, much like *Mytilus* mussels in the marine environment. Densities measured in thousands per m^2 are common [1], but tens of thousands are not rare, and one study reported as many as 114,000/m^2. The individual mussels are generally small, 1–2 cm in length, but they can grow up to 5 cm. Their biomass varies as a function of density, size, and condition and is mostly within the range of 0.2–20 kg/m^2. Their lifetime is relatively short, typically 3–8 years.

A. Jernelöv, *The Long-Term Fate of Invasive Species*,
DOI 10.1007/978-3-319-55396-2_2

Close-up of a typical shell of a zebra mussel (Photo from Wikipedia). http://en.wikipedia.org/wiki/File:Dreissena_polymorpha3.jpg

Live zebra mussels underwater with shells open, animals respiring, and siphons visible (Photo from Wikipedia)

Dependent on the size of the mussel and water temperature, a zebra mussel will filtrate 1–5 L of water a day, consuming what it finds edible and depositing feces and nonfood particles covered in mucus, so-called pseudofeces, on the bottom. Sexual maturation of zebra mussels can occur early, and cases have been reported in which, before they have been settled for 2 months, they already release eggs and sperms [2]. Normally, however, they start doing that in the second year at a size of just under a centimeter.

An adult female can produce million eggs in a year and over five million in her lifetime. Veligers, the tiny, free-swimming larvae of the zebra mussel, will settle onto any hard surface within a week or two of birth. They have some tendency to swim against the current, but are largely carried along with it [3]. Other authors have made much lower estimates of the fecundity of the zebra mussel females, giving 150,000 eggs in a year as a high.

Small zebra mussels have a number of predators in their area of origin. Fish such as sturgeon, catfish, carp, and eel feed on them, as do many species of birds and crayfish.

Copepods such as *Cyclops* also feast on veligers.

The zebra mussel is a freshwater species but tolerates salinities up to 6 PSU (or per mile). Its natural habitat includes brackish water areas of the Caspian and Black Seas.

The Spread to Western Europe

In the eighteenth century, a number of canals were dug to connect Russian rivers going to the Caspian Sea with those emptying into the Black Sea. Later, the canal system was extended further westward and connected to those in Central and Western Europe. The canals and the rivers they connected became the first steps along the route of the zebra mussel's westward spread. They were found in Hungary and Slovakia in 1794 [4], only a quarter of a century after it was first described by Pallas from specimens collected in the Ural, Volga, and Dnieper Rivers. Grossinger, at the time he found the mussel in the Danube basin, did not think of it as an invasive species, but in Rotterdam, the Netherlands, it was very much seen as such when it appeared there in 1827, particularly when it started to interfere with the operation of sluices [5].

The concept of a more western native distribution range of the zebra mussel is, however, not without modern subscribers. The December 2014 issue of the journal *Water Technology* states that "They were originally found in the Balkans, Poland and the former Soviet Union" [6]. Most scholars, however, see the Caspian and Black Sea drainage areas as the native home of the zebra mussel and some only the former.

Already prior to its detection in the Netherlands, the zebra mussel had started its conquest of the British Isles, starting in Cambridgeshire in the early 1820s, reaching London in 1824, Yorkshire in 1831–1833, the Forth and Clyde Canals in 1833, and the Union Canal, near Edinburgh, in 1834 [7, 8].

Obviously, the zebra mussel is much more likely first to have reached England on board a ship rather than attached to the exterior of one, as their saltwater tolerance probably is not high enough for them to have survived the Channel crossing otherwise. Similarly, it is likely that the mussel had reached Lake Mälaren and Sweden with ballast water, or otherwise on board a ship, when it was first spotted there a 100 years later.

By that time, the zebra mussel had long since taken over the European canal system and was common over most of Germany and France. Hamburg had been reached already in 1830, Copenhagen in 1840 [9] and the river Seine outside of Paris in 1867.

To the Mediterranean parts of Europe, the zebra mussel arrived much later. In Italy, they first appeared in Lake Garda in the north in 1973 [10] and in Tuscany in the central part in 2003 [11, 12]. In Greece, they were first reported in the early 1980s [13].

In Spain, the first populations were detected in 2001 in the Flix reservoir, from where they gradually spread until fully colonizing the Ebro basin. Four years later was found in the Sitjar reservoir on the river Mijares [14].

In the more remote areas of Northern Europe, the zebra mussel also arrived later. In the Shannon estuary in Ireland, they appeared in the 1990s, some 170 years after reaching England, and spread throughout the Shannon and Shannon-Erne catchments [15]. To Finnish territorial waters, the zebra mussel came in the 1990s, after having been present in the nearby large Russian Lake Ladoga for 150 years and in Estonia for a similarly long time [16, 17].

The invaded areas of Western Europe also include some other brackish parts of the Baltic, as well as estuaries in the North Sea and Irish Sea. Here, the densities are mostly much lower than in freshwater, and in competition with the *Mytilus* species, *Dreissena*, the zebra mussel loses out as soon as salinity is high enough to support those other marine organisms.

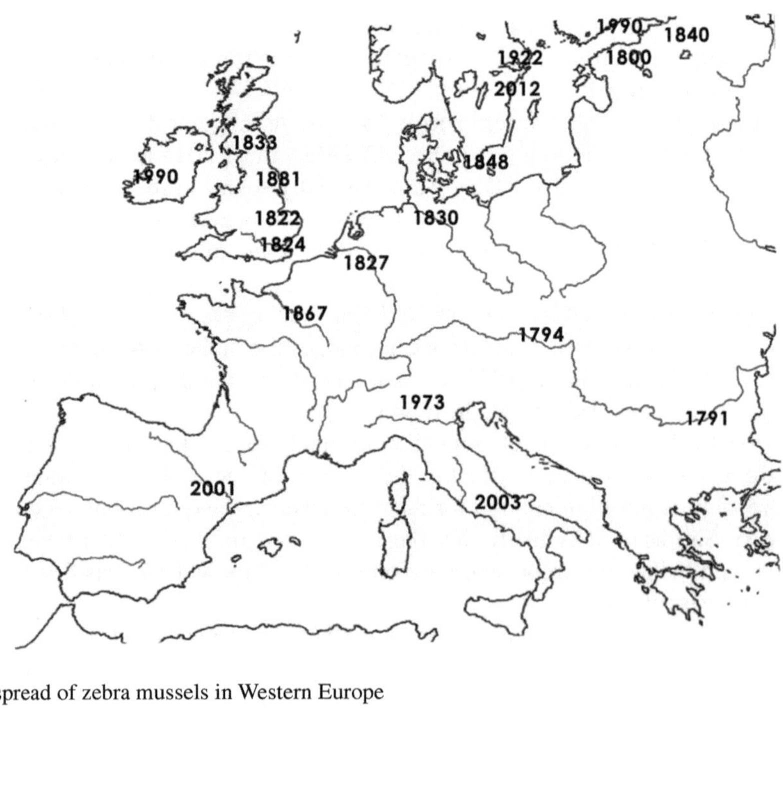

The spread of zebra mussels in Western Europe

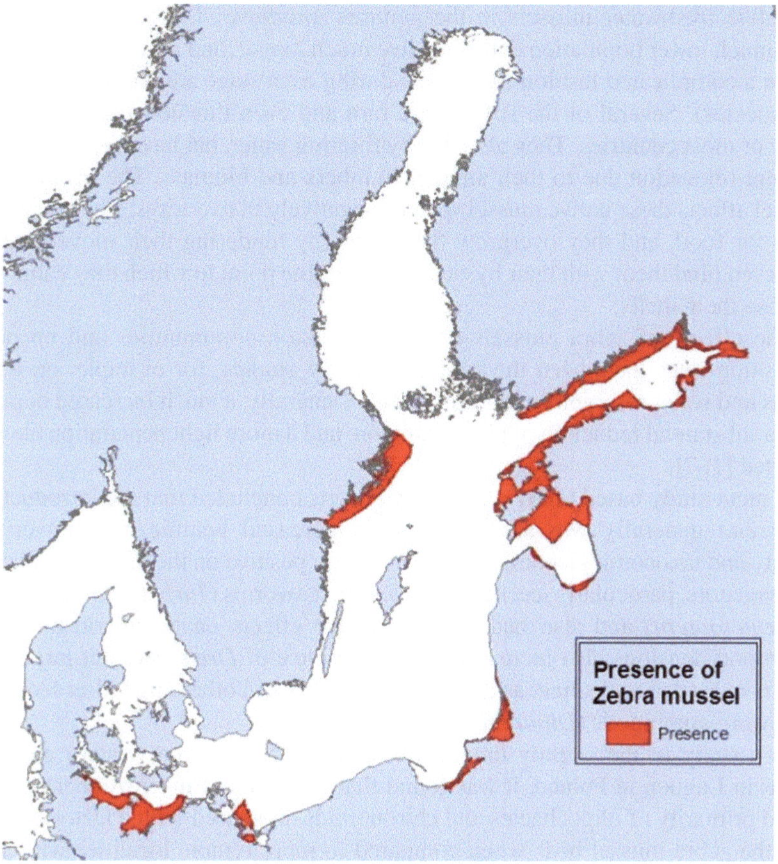

The presence of *Dreissena polymorpha* in different coastal areas in the Baltic (source: HELCOM List of non-indigenous species). Note that the distribution in the Gulf of Riga and Swedish Baltic Proper is limited to more freshwater parts

Effects of the Zebra Mussel in Newly Invaded Waters in Western Europe

In West European freshwater systems, the zebra mussel represents a new type of organism due both to the high population densities in which it occurs in suitable habitats and to its filter-feeding mechanism, which relocates nutrients from the water mass to the bottom, thereby depleting the pelagic and enriching the benthic community. Clearer water also allows sunlight to penetrate deeper, thereby extending the reach of attached algae and rooted plants.

Native freshwater mussels in the genuses *Anodonta*, *Unio*, and *Margaritifera* have much lower population densities, live much longer, and reproduce more slowly and in a complicated fashion (the larvae during each stage are parasites of different fish species). Several of the species are rare and even threatened by extinction in many or most countries. They also live by filtering water, but have little effect on the nutrient relocation due to their smaller numbers and biomass. The invading zebra mussel affects these native mussel species negatively in two ways: They outcompete them for food, and they overgrow them, thereby hindering their movements, and may even bind them with their byssus threads to the point to which they cannot open or close their shells.

The effects of zebra mussels on phytoplankton communities and on particle deposition rates have been the subject of many studies, for example, on Russian canals and water reservoirs and Polish lakes. Generally, a much increased deposition rate, a substantial reduction of phytoplankton, and a more light penetration have been reported [1, 7].

A meta-study based on 47 underlying reports concluded that the introduction of *Dreissena* generally was associated with increased benthic macroinvertebrate density and taxonomic richness. The effects were positive on the densities of scrapers and predators, particularly leeches (*Hirudinea*), flatworms (*Turbellaria*), and mayflies (*Ephemeroptera*) and also had strong positive effects on gammarid amphipods. Gastropod densities also increased in the presence of *Dreissena*, but large-bodied snail taxa tended to decline, as did sphaeriid clams and other large filter feeders and burrowing amphipods (*Diporeia* spp.) [18].

In a study in the slightly brackish waters (1 PSU) of the western part of the Szczecin Lagoon in Poland, it was found that the fauna of macrozoobenthos, comprised primarily of oligochaetes and chironomids, increased with a factor of 2.4–4.9 near the zebra mussel beds when compared to more remote localities with similar sediment structures. Chironomids, oligochaetes, and *Gammarus* amphipods were particularly favored [19].

With regard to human technical installations in water, the zebra mussel has caused significant damage. In populating supply pipes, they reduce water flows. They damage boats, engines, fishing gear, buoys, locks, and docks and turn bathing beaches into foot- and hand-cutting underwater torture fields.

After Their Introduction, What Happened Next in Western Europe?

The typical picture, repeated many times over in Russia, continental Northern Europe, Britain, Sweden, and, more recently, in Ireland and Spain, is that once established in a water system, the zebra mussel will spread throughout it. This spread will be particularly fast where boat traffic is intensive, as it is in canal systems. In this initial phase, mussel densities will also be extremely high, but after some years, densities decrease and more or less stabilizes at much lower levels [1, 20, 21].

In this context, a special double interaction between zebra mussel and crayfish has been noted. On one hand, crayfish eat (small) mussels and reduce population densities [1, 22, 23].

On the other hand, zebra mussels settle on crayfish and may impair their mobility and feeding possibilities [24].

Studies in Polish lakes have shown that roach, with their strong pharyngeal denticles, adopt to feed almost exclusively (95–100%) on zebra mussels and that their predation impact may become significant [3, 25].

Crayfish covered with zebra mussels.

Photo jpeg mcarp

When zebra mussels are present in a water system, even as natural and human-induced mechanisms for further overland spread to nearby water systems of both veligers and adult mussels can easily be imagined, these mechanisms appear to be relatively ineffective. In Sweden, for instance, the zebra mussel was introduced to Lake Mälaren in the 1920s, but some 80 years passed before they were found to inhabit another freshwater system, the Göta Kanal, despite a low-level presence in the Baltic archipelago to which both systems are connected [26].

The European attitude to the zebra mussel is not all negative. Most of Europe's inland waters have been affected by eutrophication. Increased nutrient loads, most importantly of phosphorus, have led to increased phytoplankton densities, more turbid waters, anoxic bottoms, diversity losses among aquatic plants, and a shift in fish populations. Environmental measures since the 1970s have reduced but not eliminated these problems. The presence of zebra mussels in many ways counteracts the effects of eutrophication and in some respects restores water bodies to the status they had some 100–150 years ago, which also has a positive effect on shoreline property values. This has in turn led to proposals and experimental attempts to deploy the zebra mussel for restoration of overfertilized bodies of water.

The Spread to North America and Further Invasion There

In North America, the zebra mussel was first found on the Canadian side of Lake Erie in 1988. It is assumed to have arrived with ballast water of an ocean-going vessel that entered the Great Lakes system through the Saint Lawrence Seaway. Given the size of the population and of the individual mussels, it was concluded that the introduction must have taken place some years earlier, probably in 1985 or 1986. By 1990, it had spread to all of the Great Lakes, with especially dense populations in Lake Erie and Lake St. Clair. As some river systems are connected to the Great Lakes and/or one other via canals, the zebra mussel was able quickly to expand its territory, and within less than a decade of its arrival, it was present in the Arkansas, Cumberland, Hudson, Illinois, Mississippi, Ohio, and Tennessee rivers.

Contrary to Western Europe, where zebra mussels did not easily spread from one water system to another, their progress seems to have gone faster in North America—although whether with the help of humans or as hitchhikers on other biota is a matter of sometimes intense debates. Thus, in the following years, the invaders appeared in a number of new places, e.g., in Connecticut in 2001; in a small, isolated quarry in Virginia in 2002; as veligers in the Missouri River in Nebraska and South Dakota in 2003; and so on. More recently, in 2008, they were found in the San Justo Reservoir in Central California and in Colorado, in the Pueblo Reservoir and Pueblo River and in Grand Lake, 50 miles northwest of Denver. Populations in Lake Texoma on the border of Texas and Oklahoma and in western Massachusetts were found in 2009, and zebra mussel veliger was discovered in North Dakota's Red River in 2010 [27].

By March 2014, the zebra mussel had the US distribution shown on the map below and had also entered Canadian freshwater systems in Manitoba (Lake Winnipeg), Quebec (Richelieu River), and Ontario (Lake Muskoka).

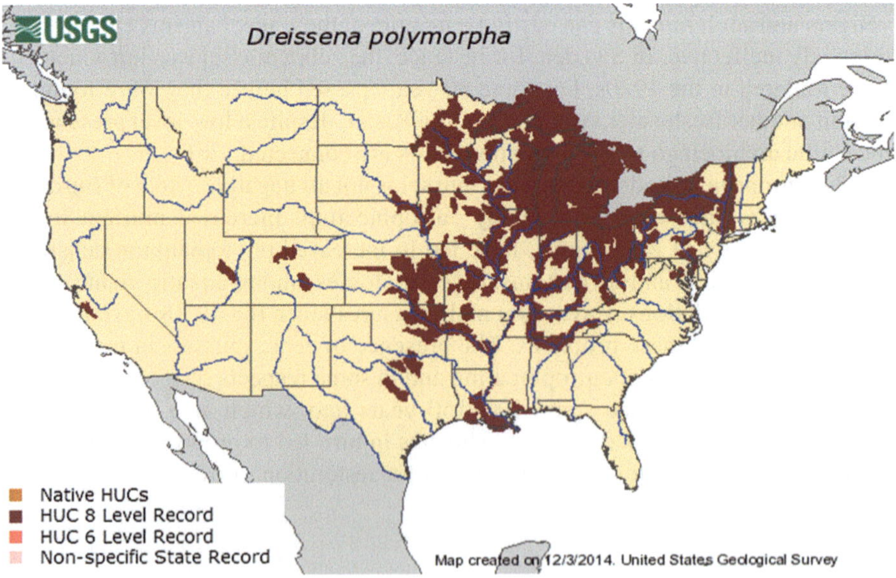

Ecological Effects of the Zebra Mussel in North America

The Great Lakes

Phytoplankton

Initial Effects

A large number of studies were performed to determine the ecological effects of the zebra mussel in the Great Lakes, particularly in Lake Erie in which the species became established in the late 1980s and where it quickly came to dominate that lake's benthic community. The first effects were captured by the many monitoring programs in place since the 1970s, when reducing nutrient loading and algal blooms first became an environmental priority. These found that Secchi disk transparency increased 100% or more and that phytoplankton densities decreased by some 90% [28].

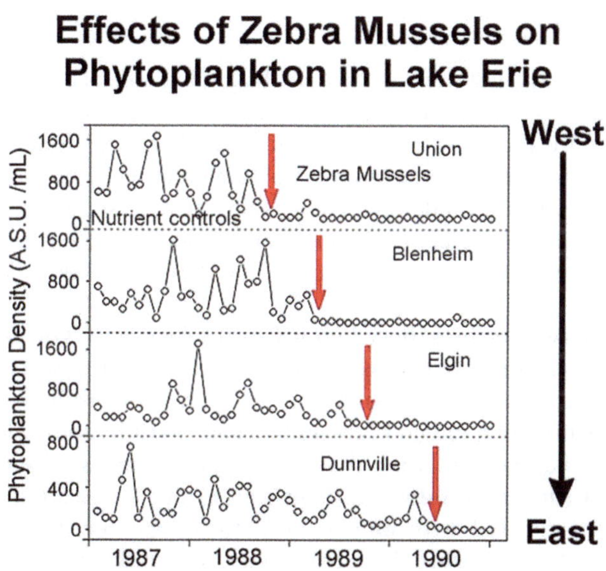

Also measured as chlorophyll A, the reduction of microscopic algae was dramatic. In the western basin (at the Union water intake), where summer concentrations averaged 6.3 μg/L during a 4-year pre-zebra mussel time period and the average value declined to 1 μg/L during a 4-year post-zebra mussel time period. Total phosphorus and total nitrogen also went down, phosphorus by the most, however, which resulted in a doubling of summer total N to total P ratios after the arrival of zebra mussels [29]. That report was based on monitoring data from Dr. Ruth Holland Beeton. So another report using data from an area near the Stone Laboratory on Lake Erie said that water transparency was about a meter in the 1970s, before phosphorus abatement programs began, and improved to 2–3 m in the 1980s after a decade of reduced phosphorus inputs. Water transparency continued to improve, to 3–6 m, after zebra mussels colonized the area [30].

A study quantifying phytoplankton production and zebra mussel algae removal in Lake Erie in 1990 found that the zebra mussel population consumed 5.0 million tons of phytoplankton, while a further 1.4 million tons of phytoplankton were captured and deposited as pseudofeces. Thus, a total of 6.4 million tons of phytoplankton were removed from the water column by zebra mussel in western Lake Erie during that year. As primary production was estimated to be 24.8 million tons, zebra mussel removed the equivalent of 26% of the primary production for western Lake Erie [31].

The exact figures differed between studies and between the lakes and lake basins in the Great Lakes region, but overall, the pattern was the same—a dramatic decrease in planktonic algae and increased water clarity following the introduction of the zebra mussel.

Subsequent Development

From the mid-1990s, the trend of low algal standing stocks during summer was reversed with the appearance of toxic *Microcystis* blooms. A large study, supported in part by the Lake Erie Protection Fund, was undertaken to explore hypothesized mechanisms of *Microcystis* bloom creation by zebra mussels using statistical and dynamic modeling. The ensuing results supported the idea that zebra mussels, through selective feeding and nutrient excretion and by avoiding the toxic *Microcystis*, promoted blooms in both Saginaw Bay, Lake Huron, and Lake Erie [32].

I have some reservations, based on my own experience with zebra mussels, about these conclusions. In connection with a project to test whether the zebra mussel could be used to treat water from fish farms before its release, we also tested how the mussel handled blue-green algae. So did Vanderploeg et al. [33] and we found, like they did, that (most of) the cyanophytes were not consumed but excreted with the pseudofeces. However, in our tests, the algae were not unharmed, but covered with mucus, and stuck together in sediment-bound lumps, effectively preventing them from returning to the water mass. Further, they and we found that nutrient recycling from zebra mussels increased the N/P ratio (in line with the Lake Erie results in [29] above). As blue-green algae with nitrogen fixation abilities are especially competitive when phosphorus levels are high and nitrogen levels are low, this result shows, in my understanding, it to be less, rather than more, likely that a *Microcystis* bloom is induced when zebra mussels are present in high numbers.

Zooplankton

The zooplankton community has also been affected by the invasion of zebra mussels. Zooplankton abundance dropped 55–71% following mussel invasion in Lake Erie, with microzooplankton more heavily impacted. Mean summer biomass of zooplankton decreased from 130 to 78 mg dry wt. m-3 between 1991 and 1992 in the inner portion of Saginaw Bay and Lake Huron [34].

Benthic Fauna

Initial Effect

The bottom fauna represented the mirror image of the planktonic community. In a comparison between two studies undertaken with the same methodology, in the same area of western Lake Ontario, one pre- and one post-*Dreissena* found that in 1991–1992, the zebra mussel accounted for 93% of the macroinvertebrates (insects, crustaceans, worms, and mollusks) collected, replacing the sideswimmer (*Gammarus fasciatus*) as the numerically dominant taxon, a place the amphipod held in 1983. Although the number of sideswimmers was also significantly higher, the sheer number of non-*Dreissena* macroinvertebrates was many times greater in 1991–1992 than in 1983, skyrocketing from 127–1159/m^2 to 1316–5267/m^2. Other taxa with great increases in numbers were annelid worms (*Manayunkia speciosa, Spirosperma ferox*) and unidentified tubificids, gastropod snails (*Helisoma anceps, Physa heterostropha, Stagnicola catascopium, Valvata tricarinata, Goniobasis livescens, Amnicola limosa*), insects of the order *Trichoptera* (*Polycentropus* sp.), and the decapod crayfish (*Orconectes propinquus*). No taxon was less abundant in 1991–1992 than in 1983 [35, 36].

The study concluded that:

> "by occurring in large clumps and filter-feeding intensively, Dreissena may benefit other macroinvertebrate taxa in Lake Ontario by increasing the complexity of benthic substrate and by increasing the flow of energy to benthic environments. While some factors (e.g., increased water clarity due to Dreissena biofiltration and phosphorus abatement programs) may be acting synergistically to favorably impact benthic organisms, biodeposition by Dreissena may keep benthic food resources at high levels despite overall declines in fertility of the Lake Ontario ecosystem." [37]

Similar population changes were reported by scholars who studied the impacts of the zebra mussel on benthic macroinvertebrates in Lake Eire and Lake St. Clair at about the same time.

A group of benthic organisms that suffered heavily from the zebra mussel invasion, however, was the native unionid mussel. Zebra mussels readily colonize clam shells, disrupting feeding, movement, and reproduction. The otherwise long-lived clams generally die within 1–2 years after infestation. This die-off has been well documented in the Great Lakes [38, 39], with near total mortality reported throughout most of western Lake Erie.

Many zebra mussels attached to a native mussel (Photo from Minnesota Department of Natural Resources)

Subsequent Development

However, in 1996, a large population of native clams was discovered in a western Lake Erie wetland, Metzger Marsh [37], and subsequently more populations were detected, mostly in shallow waters in which the zebra mussel does not thrive. In an interview, one of the active scientists, Dr. Dave Zanatta, gave an account of the development timeline:

> "Native mussels were already threatened by pollution. Once zebra mussels arrived, they were nearly extirpated. Come the late 1980s and early 1990s, all the native mussel populations crashed in Lake St. Clair, Lake Erie and the Great Lakes in general. Average density of native mussels before the arrival of zebra mussels was two per square meter in Lake St. Clair. By 1990, zebra mussel density was at 1,600 per square meter.
>
> By 1992, native mussel populations are almost gone from the southeastern portion of the lake and declining rapidly in the northwestern portion of the lake. By 1994, there were almost no native mussels left in the lake, with zebra mussels now at 3,000+ per square meter. But there was reason for hope. Remnant populations of native mussels were beginning to be found in coastal wetlands in western Lake Erie in the late 1990s. In addition, zebra mussel populations started showing dramatic declines between 1994 and 2001 in Lake St. Clair, even though their sheer numbers remained staggering."

> "In expanded research funded through the Great Lakes Restoration Initiative, we found that the St. Clair Delta and western Lake Erie were the most healthy areas of the lakes in terms of native mussel abundances. Most of the 124 sites he investigated contained more than five unique species of native mussel. The native mussels appear to prefer very shallow water, less than two meters deep, with soft sediments, fluctuating water levels, good wave action and the fast water flow in the Delta, all of which curb the spread of zebra mussels and help make them available to predators." [40, 41]

Fishes

There are a large number of articles postulating that fish that largely feed in the pelagic zone, such as the walleye, will suffer, both as the organisms on which they feed are depleted and as their spawning grounds are taken over by zebra mussels. There is, however, hardly any factual data that show such a decrease. In fact, the spawning ground argument has been contradicted by observations in the Great Lakes of walleye successfully spawning over zebra mussel beds [42] and evidence that the peak of the walleye population in the period 1979–2005 partly overlapped with the peak density of zebra mussels.

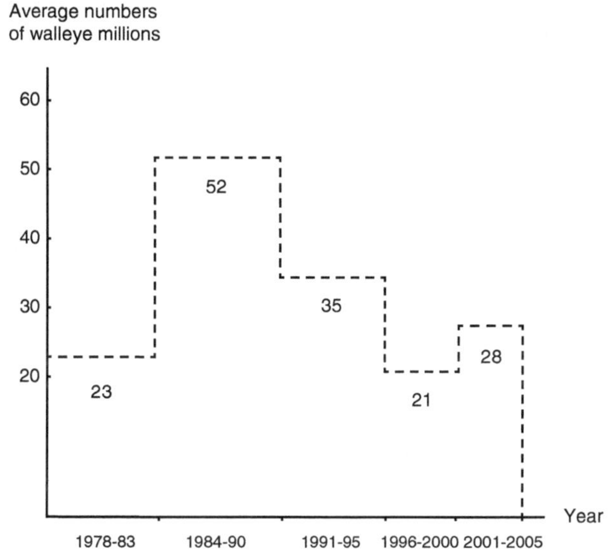

Walleye population (2+) in Lake Erie from 1978 to 2005. Based on data from the Detroit River-Western Lake Erie Basin Indicator Project [43]

Other fish species, such as smallmouth bass and yellow perch that rely more on benthic organisms for food, have in line with forecasts benefitted from the bottom fauna enrichment provided by the zebra mussel and have also gradually increased consumption of the mussel itself as part of their diets. In Lake St. Clair, the yellow perch population rose in the late 1980s and 1990s, as did the smallmouth bass population of Lake Eire [44].

In the Huron River, the increase in the smallmouth bass has been attributed to the increase in benthic macrophytes, following increased water transparency, providing suitable places for spawning and for fry growth and development [45].

Lake sturgeon, freshwater drum, catfish, and all sunfish species also feed heavily on zebra mussels. Some migratory ducks have even changed their migration stopover habits so they can feast for some days on the zebra mussel beds [46].

The Quagga Mussel, Dreissena rostriformis bugensis

The quagga mussel is a close relative of the zebra mussel. Quagga was the name of a subspecies of African plains zebra, extinct since the nineteenth century, with pale brown and white stripes mainly on the front of the body and solid brown on the rump. The quagga mussel was named precisely for this pattern of colorization (although those in Lake Erie are pale or white and have no stripes) and its close relation to the zebra mussel. The quagga mussel came from the Dnieper River drainage area in Ukraine, an area neighboring that of the zebra mussel, and in most respects, the two species are quite similar.

The quagga mussel came to the Great Lakes through the same route as the zebra mussel, namely, in ballast water. In Lake Erie, it was first collected in 1989, near Port Colborne, Ontario, but was not identified as a different species from the zebra mussel until 2 years later.

Like the zebra mussel, the quagga mussel spread throughout the Great Lakes system and beyond, but the spread was slower, and it has not yet reached the same distribution.

On the other hand, in the part of the Great Lakes, notably Lake Erie and Lake St. Clair, to which the two *Dreissena* species first came, the quagga mussel has now taken over and totally dominates: a study in Lake Erie from 2004 revealed a drop in total *Dreissena* density with 50% compared to 4 years earlier, to some 4000 individuals/m^2, but noted that about 97% were quagga mussels [47].

The ecological and economic effects of the two species are very similar.

Also in Western Europe, the quagga mussel has followed in the footsteps of its zebra mussel cousin, but it started a century later. Only in the 1940s did it start the spread from its Dnieper River native area, and up to the 1980s, the westward advance was slow. Since then, the speed has been higher and is now present in a large part of the waters of Continental Europe. Once in a water system, the zebra mussel quickly colonizes it and reaches the peak density within 1–4 years. For the quagga mussel, it takes 6–19 years [48].

In the Dnieper River basin, where the species coexist, the dominance shifts between areas and years. In water bodies invaded by both species, *D. polymorpha* generally has the upper hand in those with no or only a small profundal zone, especially in the shallow areas, while *D. rostriformis bugensis*, after a decade or more, dominates where the profundal zone is large.

Hudson River

The zebra mussel was first found in the Hudson River estuary in May 1991. The population grew extremely rapidly and had reached an average density over the freshwater tidal river of 4000/m^2 within 18 months. With an estuarine surface area of 135 km^2, this means a staggering 550 billion individuals. They constituted more than 70% of the zoobenthic biomass and almost exclusively lived on sub-tidal rocks.

After the rapid population growth in 1991–1992, their reproductive success fell four orders of magnitude in 1993–1994, and the young of the year were small, only 20–30% of the size of those in 1991 [49]. In all likelihood, the adult zebra mussels outcompeted the larvae for food or just cannibalized a major part of the veligers.

Phytoplankton biomass declined 85% following mussel invasion in the Hudson River [50]. The total biomass of zooplankton declined 70%, due both to a reduction in body size of large zooplankton and a reduction in abundance of microzooplankton, the first as a result of less phytoplankton to feed on and the latter from direct predation by the zebra mussels [51].

A decade later, the situation was very different. The number of mussels was still very large, but their dry biomass (without shells) had gone from 3000 tons to a small fraction thereof, as there were few individuals over a year old. Since the filtering capacity of adolescents, both with regard to volume and to particle size, is much smaller than of adults, this allowed zooplankton such as *rotifers* to return to almost pre-zebra mussel invasion densities of about 1000 individuals per liter in the summer. The main reason for this dramatic change in the zebra mussel population was that blue crabs (*Callinectes sapidus*) that migrate into the estuary in the late summer consumed the adult zebra mussels and over the time period of a decade did so to such an extent that only a small proportion of the mussels survived. Only the high fecundity of the zebra mussel kept their numbers up. As the blue crabs themselves had not increased notably in numbers during this time, the most likely explanation is that zebra mussels had become a gradually larger part of their diet, although it could not be ruled out that other, larger predators also contributed to the zebra mussel consumption. The process can be seen as one in which a native predator with time "tames" an invasive species [52].

Economic Effects of the Zebra Mussel in North America

Solid structures, in particular those of steel or concrete standing in running water with relatively low and even flow rates, are ideal places for the zebra mussel to settle. Consequently, many man-made structures attract large conglomerations of *Dreissena* mussels. There, the mussels quickly multiply, enough significantly to reduce flow rates. The zebra mussel thus colonizes water supply pipes of hydroelectric and nuclear power plants, public water supply plants, and industrial facilities. It populates the pipes' interiors, constricting water flow and reducing the intake in heat exchangers, condensers, firefighting equipment, and air-conditioning and cooling systems. Zebra mussels can also damage boats, engines, navigational buoys, fishing gear, and dock pilings, as well as diminish steel and concrete structural integrity. In the USA, congressional researchers estimated the mussel costs the power industry alone $3.1 billion in the 1993–1999 period, with its combined impact on industries, businesses, and communities climbing to over $5 billion [53] Since then, the zebra mussel population has almost collapsed in the areas into which it was first introduced, but has spread to new ones. The annual control costs are likely to be smaller, although some sources project an estimated cost of $5 billion for the coming decade [54].

The intensity of the problem has decreased in the area around the Great Lakes, where it first emerged, but it has spread geographically. Technical adjustments, such as using Teflon-like material on which the mussel does not so easily settle, have also reduced the costs for regular mussel removal, but at the price of higher investment costs.

There are no methods available to kill or remove zebra mussels on a large scale in the environment. In technical installations, mechanical removal and killing the adolescent mussels with hot water, chlorine gas, salt, or detergents are methods in use. To overcome the rejection and valve-closing response generally seen when zebra mussels are exposed to toxic substances, more advanced methods of delivery are sometimes practiced. Foremost among these is microencapsulation of toxins in particles that are edible for the mussels. The active ingredient used is potassium chloride, which is not lethal at low doses to most organisms, including fish, but which is particularly toxic to freshwater bivalves. Another emerging control for *D. polymorpha* is the use of endocannabinoids, anandamide, and other compounds that have been tested to inhibit zebra mussel byssal attachment. These naturally occurring and synthetic cannabinoids can serve as efficacious, nontoxic zebra mussel anti-foulants [55].

Some attempts have also been made with biological control agents such as the common soil bacterium, *Pseudomonas fluorescens*, which is toxic to the zebra mussel but harmless to man. No such methods have yet reached commercial applications.

In some situations, in which *Dreissena* mussels destroy bathing beaches or hinder operations of recreational boats, their presence has reduced property values. The bigger picture, however, is that clearer and more transparent water increases beachfront property values [56]. No large-scale, nationwide assessment of this economic effect seems to have been done yet, though.

The Likely Future

Based on experience in Western Europe, it seems likely that the spread of the zebra mussel to new water systems in the suitable temperature ranges of North America will continue and that following their introduction, rapid colonization, and mass development, the associated ecological effects and economic problems will occur there, too.

In areas in which the mussel is already present in the initial mass occurrence, the densities are likely to go down and the ecological effects to become milder. However, the need to prevent them from settling or to remove mussels from technical installations will probably remain high. The fecundity of the zebra mussel means that even with fewer adults around, there will still be plenty of veligers eager to settle. In addition, many man-made surfaces, for example, water intake pipes, provide not only a suitable hard substrate and a water stream with food and oxygen but also protection from predators.

In Europe, the slow advance in the Mediterranean area is likely to continue, and with a warming climate, the zebra mussel may go on to conquer more northern territory as well.

As the slower spread of the quagga mussel also continues, the two *Dreissena* species are likely to share territory and split dominance dependent on factors such as temperature, depth, sediment structure, and salinity. In areas where the zebra mussel is the first and only invader now, this means that it is likely that it will eventually have to cede some territory to its cousin.

Summary

Since the first part of the nineteenth century, the zebra mussel has invaded most freshwaters of Western Europe, as well as, since the 1980s, important parts of the waterways of the Northeastern USA and southern Canada. It is an extremely efficient invader and often, within a few years after having entered a water system, totally dominates the non-photosynthesizing biomass. They will then filter and clear large volumes of water, not seldom a whole lake within days or a week, consuming the phyto- and micro-zooplankton and extracting calcium to build their shells. As a result, the pelagic part of the food web becomes impoverished, accompanied by a corresponding flourishing of the littoral part. However, other mussel species there may suffer severely from the competition. In many ways, the impacts of the zebra mussel on the aquatic ecosystem are the opposite of those of eutrophication. Mostly, the period of extreme zebra mussel population density is short, typically a decade or so, after which time the native predators have learned to consume the newcomers and their densities decrease with a factor of 10 or more.

A number of man's installations in water, including water intake pipes, offer attractive settling places for the zebra mussel and necessitate costly removal operations.

Generally speaking, it seems that both the ecological and economic effects of zebra mussel invasions have been more pronounced in North America than in Europe. It could, however, be that more attention is now paid to ecological effects than was the case 150 years ago and also that water-based technical systems are much more prolific and sensitive now than they were then.

References

1. Stanczykowska, A. (1977). Ecology of *Dreissena polymorpha* in lakes. *Polish Archives of Hydrobiology, 24*(4), 461–530.
2. Borcherding, J. (1991). The annual reproductive cycle of the freshwater mussel *Dreissena polymorpha* Pallas in lakes. *Oecologia, 87*, 208–218.

3. Stanczykowska, A., & Lewandowski, K. (1993). Thirty years of studies of *Dreissena polymorpha* in Masurian Lakes of northeastern Poland. In T. F. Nalepa & D. W. Schloesser (Eds.), *Zebra mussels: Biology impacts and control* (pp. 3–33). Boca Raton, FL: Lewis Publishers.
4. Grossinger, J. B. (1794–1797). *Universa historia physica regni Hungariae secundum tria regna naturae digesta*. Tomus I–V. Posonii et Comaromii.
5. Bij de Vaate, A., Jazdzewski, K., Ketelaars, H. A. M., Gollasch, S., & Van der Velde, G. (2002). Geographical patterns in range extension of Ponto-Caspian macroinvertebrate species in Europe. *Canadian Journal of Fisheries and Aquatic Sciences, 59*, 1159–1174.
6. Retrieved December 2, 2014, from http://www.watertechonline.com/articles/zebra-mussels
7. Mackie, G., Gibbons, W., Muncaster, B., & Gray, I. (1989). *The zebra mussel, Dreissena polymorpha: A synthesis of European experiences and a preview for North America*. Ontario Ministry of Environment.
8. Aldridge, D. C. (2010). Chapter 7: *Dreissena polymorpha* in great Britain: History of spread, impacts and control. In G. Van der Velde, S. Rajogopal, & A. bij de Vaate (Eds.), *The zebra mussel in Europe* (490 pp). Leiden: Backhuys Publishers.
9. Morton, B. S. (1969). Studies on the biology of *Dreissena polymorpha* 4. Habits, habitats, distribution and control. *Water Treatment and Examination, 18*, 233–241.
10. Giusti, F., & Oppi, E. (1973). *Dreissena polymorpha* nuovamente in Italia. *Memorie del Museo Civico di Storia Naturale di Verona, 20*, 45–49.
11. Lori, E., & Cianfanelli, S. (2006). New records of *Dreissena polymorpha* from Central Italy. *Aquatic Invasions, 1*(4), 281–283.
12. Cianfanelli, S., Lori, E., & Bodon, M. (2010). Chapter 8: *Dreissena polymorpha*: Current status of knowledge about the distribution in Italy. In G. Van der Velde, S. Rajogopal, & A. bij de Vaate (Eds.), *The zebra mussel in Europe*. Leiden: Backhuys Publishers. 490 pp.
13. Morton, B. (2000). *Chapter 1: The aquatic nuisance species problem: A global perspective and review* (pp. 1–54). Lewis Publishers.
14. Lalaguna, C. D., & Marco, A. A. (2008). The zebra mussel invasion in Spain and navigation rules. *Aquatic Invasions, 3*(3), 315–324.
15. Retrieved December 3, 2014, from http://www.fisheriesireland.ie/Invasive-species-list/zebra-mussel.html
16. Birnbaum, C. (2011). NOBANIS—Invasive Alien Species Fact Sheet—*Dreissena polymorpha*.—From: Online Database of the European Network on Invasive Alien Species. Date of Access December 3, 2014.
17. Valovirta, I., & Porkka, M. (1996). The distribution and abundance of *Dreissena polymorpha* (Pallas) in the eastern gulf of Finland. *Memoranda Societatis pro Fauna et Flora Fennica, 72*, 63–78.
18. Ward, J. M., & Ricciardi, A. (2007). Impacts of Dreissena invasions on benthic macroinvertebrate communities: A meta-analysis. *Diversity and Distributions, 13*, 155–165.
19. Radziejewska, T., Fenske, C., Wawrzyniak-Wydrowska, B., Riel, P., Wozniczka, A., & Gruszka, P. (2009). The zebra mussel (*Dreissena polymorpha*) and the benthic community in a coastal Baltic lagoon: Another example of enhancement? *Marine Ecology, 30*(Suppl. 1), 138–150.
20. Walz, N. (1973). Studies of the biology of *Dreissena polymorpha* in Lake Constance. *Archiv für Hydrobiologie*, (Suppl. 42), 452–482.
21. Stanczykowska, A., Lewandowski, K., & Czarnoleski, M. (2010). Chapter 11: Distribution and densities of *Dreissena polymorpha* in Poland—Past and present. In G. Van der Velde, S. Rajogopal, & A. bij de Vaate (Eds.), *The zebra mussel in Europe* (490p). Leiden: Backhuys Publishers.
22. Piesik, Z. The role of the crayfish *Orconectes limosus* in extinction of *Dreissena polymorpha* subsisting on steelon-net. *Polski Archiwum Hydrobiologii, 21*(3–4), 401–410.
23. Reynolds, J. D., & Donohoe, R. (2001). Crayfish predation experiment on the introduced zebra mussel, *Dreissena polymorpha*, in Ireland, and their potential for biocontrol. *Bulletin Francais de la Peche et de la Pisciculture, 361*, 669–681.
24. Stucki, T. P. (1999). Life cycles and life history of *Astacus leptodactylus* in Chalzensee Pond (Zurich) and Lake Ageri., Switzerland. *Freshwater Crayfish, 12*, 430–448.

25. Mackie, G. L., & Schloesser, D. W. (1996). Comparative biology of zebra mussels in Europe and North America: An overview. *American Zoologist, 36*, 244–258.
26. Retrieved December 5, 2014, from http://www.nrm.se/faktaomnaturenochrymden/djur/ryggradslosadjur/snackorochmusslor/svenskasotvattensmusslor/vandrarmussla.1785.html
27. Retrieved December 7, 2014, from http://nas.er.usgs.gov/queries/FactSheet.aspx?speciesID=5
28. Holland, R. E. (1993). Changes in planktonic diatoms and water transparency in Hatchery Bay, Bass Island Area, Western Lake Erie since the establishment of the zebra mussel. *Journal of Great Lakes Research, 19*(3), 617–624.
29. Nicholls, K. H., & Standke, S. J. (1997). Effects of zebra mussels on chlorophyll, nitrogen, phosphorus and silica in north shore waters of Lake Erie. In. M. Munawar (Ed.), *State of the Lake Erie Ecosystem (SOLE): Past, present and future* (37p). Ont. Min. Envir. Energy Tech. Report.
30. Retrieved December 6, 2014, from http://ohioseagrant.osu.edu/_documents/publications/FS/FS-045%20Zebra%20mussels%20in%20North%20America.pdf
31. Madenjian, C. P. (1995). Removal of algae by the zebra mussel (*Dreissena polymorpha*) population in western Lake Erie: A bioenergetics approach. *Canadian Journal of Fisheries and Aquatic Sciences, 52*(2), 381–390.
32. Vanderploeg, H. A., Liebig, J. R., Carmichael, W. W., Agy, M. A., Johengen, T. H., Fahnenstiel, G. L., & Nalepa, T. F. (2001). Zebra mussel (*Dreissena polymorpha*) selective filtration promoted toxic *Microcystis* blooms in Saginaw Bay (Lake Huron) and Lake Erie. *Canadian Journal of Fisheries and Aquatic Sciences, 58*, 1208–1221.
33. Vanderploeg, H. A., Wilson, A. E., Johengen, T. H., Bressie, J. D., Sarnelle, O., Liebig, J. R., Robinson, S. D., & Horst, G. P. (2013). Chapter 32: Role of selective grazing by dreissenid mussels in promoting toxic Microcystis blooms and other changes in phytoplankton composition in the Great Lakes. In *Quagga and zebra mussels: Biology, impacts, and control*. Boca Raton, FL: CRC Press. Retrieved January 20, 2017, from https://www.glsc.usgs.gov/publications/2013/1761
34. MacIsaac, H. J., Lonnee, C. J., & Leach, J. H. (1995). Suppression of microzooplankton by zebra mussels: Importance of mussel size. *Freshwater Biology, 34*, 379–387.
35. Bader, A. P. (1985). *Dynamics of benthic macroinvertebrates inhabiting an artificial reef and surrounding areas in southwestern Lake Ontario*. M.S. thesis, SUNY College of Environmental Science and Forestry, Syracuse, NY.
36. Steward, T. W., & Haynes, J. M. (1994). Benthic macroinvertebrate community of southwestern Lake Ontario following invasion of Dreissena. *Journal of Great Lakes Research, 19*(2), 479–493.
37. Retrieved from http://www.eng.buffalo.edu/glp/articles/glrr_v3n1_1997.html
38. Schloesser, D., & Nalepa, T. (1994). Dramatic decline of unionid bivalves in offshore waters of western Lake Erie after the invasion by the zebra mussel, *Dreissena polymorpha*. *Canadian Journal of Fisheries and Aquatic Sciences, 51*, 2234–2242.
39. Schloesser, D., Nalepa, T., & Mackie, G. (1996). Zebra mussel infestation of unionid bivalves (Unionidae) in North America. *American Zoologist, 36*, 300–310.
40. *Written off, doomed native mussels return*. Retrieved from http://voicenews.com/articles/2014/10/09/news/doc542d9541162c0893321344.txt?viewmode=fullstory
41. Zanatta, D. T., Bossenbroek, J. M., Burlakova, L. E., Crail, T. D., de Szalay, F., Griffith, T. A., Kapusinski, D., Karateyev, A. Y., Krebs, R. A., Meyer, E. S., Paterson, W. L., Prescott, T. J., Rowe, M. T., Schloesser, D. W., & Walsh, M. C. (2015). Distribution of native mussel (Unionidae) assemblages in coastal Lake Erie, Lake St. Clair, and connecting channels, twenty-five years after the dreissenid invasion. *Northeastern Naturalist, 22*(1), 223–235.
42. Fitzsimons, J. D., Leach, J. H., Nepszy, S. J., & Cains, V. W. (1995). Impacts of zebra mussel on walleye (*Stizostedion vitreum*) reproduction in western Lake Erie. *Canadian Journal of Fisheries and Aquatic Sciences, 52*(3), 578–586.
43. Retrieved December 9, 2014, from http://www.epa.gov/med/grosseile_site/indicators/walleye.html
44. Retrieved December 9, 2014, from http://seapics.com/feature-subject/marine-invertebrates/zebra-mussel-pictures.html

45. Retrieved December 9, 2014, from http://ohioseagrant.osu.edu/discuss/index.php/topic,408.0.html
46. USGS. *Science for a changing world.* Zebra mussel mshelgeseg.pbworks.com/w/file/fetch/60724393/Zebra%20Mussel
47. Culver, D. A., & Conroy, J. D. (2007). *Impact of dreissenid mussel population changes on Lake Erie nutrient dynamics* (A Final Report to the Ohio Lake Erie Commission). Ohio Lake Erie Protection Fund. Project SG 233–04.
48. Karatayev, A. Y., Burlakova, L. E., Mastitsky, S. E., Padilla, D. K., & Mills, E. L. (2011). Contrasting rates of spread of two congeners, *Dreissena polymorpha* and Dreissena Rostriformis Bugensis, at different spatial scales. *Journal of Shellfish Research, 30*(3), 923–931.
49. Strayer, D. L., Powell, J., Ambrose, P., Pace, M. L., & Fischer, D. T. (2011). Arrival, spread, and early dynamics of a zebra mussel (*Dreissena polymorpha*) population in the Hudson River estuary. *Canadian Journal of Fisheries and Aquatic Sciences, 53*(5), 1143–1149.
50. Caraco, N. F., Cole, J. J., Raymond, P. A., Strayer, D. L., Pace, M. L., Findlay, S. E. G., & Fischer, D. T. (1997). Zebra mussel invasion in a large, turbid river: Phytoplankton response to increased grazing. *Ecology, 78*, 588–602.
51. Retrieved from http://nas.er.usgs.gov/queries/FactSheet.aspx?speciesID=5
52. Carlsson, N. O. L., Bustamante, H., Strayer, D. L., & Pace, M. L. (2011). Biotic resistance on the increase: Native predators structure invasive zebra mussel populations. *Freshwater Biology, 56*, 1630–1637.
53. Retrieved December 2, 2014, from http://el.erdc.usace.army.mil/zebra/zmis/zmishelp/economic_impacts_of_zebra_mussel_infestation.htm
54. Retrieved December 11, 2014, from http://www.mtcws.mtu.edu/Education/2009_Posters/VanDerLaar_Policy%20EC%20Poster.pdf
55. Retrieved December 11, 2014, from http://www.issg.org/database/species/ecology.asp?si=50
56. Retrieved December 2, 2014, from http://www.jsonline.com/news/wisconsin/property-values-rise-on-zebra-mussel-infested-lakes-study-shows-b99224285z1-250495531.html

Canadian Waterweed in Europe

> *"the fearsome day is close at hand, when all freshwaters of Europe will be full of this weed, so that no ship can sail, no man can bathe, no duck can dive and no fish can swim."*

Some Basic Plant Facts

Canadian waterweed (*Elodea canadensis*) is a perennial, submerged, aquatic plant, native to most of North America. It also goes under the popular name of Canadian pondweed. The plant has both an extensive vegetative reproduction, where even the smallest fraction of a leaf or stem can grow into a whole new plant, and a sexual one with flowers, with separate male and female plants. The plant has roots, and the first shoots in the spring are often rooted in sediments. Later in the summer, though, especially when numerous free-floating loops grow out from plant fragments suspended in the water, their roots often do not even touch the bottom.

A. Jernelöv, *The Long-Term Fate of Invasive Species*,
DOI 10.1007/978-3-319-55396-2_3

Elodea canadensis (Photo from Wikipedia)

Introduction and Spread in Europe

In Europe, Canadian waterweed was first seen in Ireland, where its discovery in 1836 by Mr. J. New, in a pond outside Waringstown, is the earliest verifiable find [1]. There are reports of earlier sightings, but these have not been substantiated. The Canadian waterweed then quickly spreads over the Emerald Isle, causing substantial problems for boat traffic on narrow waterways and for the operation of locks. In did not take long for it to reach Great Britain. There, too, one can find unsubstantiated reports of its presence earlier in other localities, but the first verifiable discovery was made by Dr. G. Johnston, in a small lake at Duns Castle in Scotland.

The story of the invasion of ponds, lakes, rivers, and canals in Scotland, England, and Wales was very similar to that in Ireland. A few years after the first reported occurrence, the waterweed could be found in such masses as to be an obstacle not only to fishing and recreational use of the affected watercourses but also to navigation, and it quickly came to be nicknamed "the water pest." Later, translated to, for instance, German and the Scandinavian languages, this would become the official popular name of *Elodea* in Northern and Central Europe.

The general belief is that *Elodea* came to Ireland with wet timber from North America. This mode of dispersion between European localities was later demonstrated both in Sweden and in Russia. The fact, however, that Canadian waterweed plants can grow from very tiny fragments of leaves or stems which, in turn, can easily be transported—by boats, on fishing gear, and by aquatic birds, to name a few methods—means that there are many ways for it to spread. And spread it did, also often with direct human help.

In 1859, an intentional release from the Berlin Botanical Garden into ditches around Charlottenhof is generally regarded as the start of the rapid spread of the species in the rivers of Havel and Oder and their canal systems. This might well have been the case, but there are verified reports of *Elodea* in Lake Tegel, also in Berlin, even before that date [2].

That the Berlin releases were the only ones that triggered the spread of the Canadian waterweed in Continental Europe is questionable. The discovery of *Elodea* in Belgium in the summer of 1860 in a stopover area for migratory aquatic birds may indicate a direct route from Britain.

From Germany, the plant spread north to Denmark, where it was first reported in 1870, and to Sweden. In a pond, Brogårdsdammen, near the town of Skara in the southwest of Sweden, massive amounts of Canadian waterweed were harvested in 1873 and 1874. In an article published in 1878 describing the experience, J. Eriksson wrote: "10 full cartloads had grown during the six weeks since the pond was last cleaned out, and this, despite the fact that the weather had been fairly cold, so that the water several times had been ice-covered, and despite the fact that the water surface was no larger than about an acre" [3].

Although this was the first report, it was not the first occurrence of *Elodea* in Sweden. In collections made in 1871 from waters around Uppsala, the species was later identified [4].

The Canadian waterweed also spread east, creating the same types of problems in Poland, where it arrived in 1877, as it had done a decade earlier in Germany. In 1880, it was first spotted in St. Petersburg. By that time, it had also reached several rivers and canals in European Russia, showing the same explosive development almost everywhere.

Mass occurrence of *Elodea canadensis* in Dąbie Lake near Szczecin (NW Poland). Photo: Krzysztof Ziarnek (Permission is granted to copy, distribute, and/or modify this picture under the terms of the GNU Free Documentation License)

The waterweed colonized the Dnieper River in the 1890s and the Volga in the 1900s. It came to Southern Finland in 1905 and to Arkhangelsk, in the northern part of European Russia, in 1935. In Eastern Europe, botanical gardens and the aquarium trade are seen as major vehicles of distribution [5].

Ecological Effects of Dense Stands of Canadian Waterweed

Most of the complaints during the period of explosive growth and spread of the Canadian waterweed in the nineteenth century in Europe had to do with the mere physical effects of the masses of plants—obstacles to navigation, fishing, swimming, etc.—but there were also a number of ecological effects associated with the mass development of this aquatic plant. One was that they grew so densely that they reduced light penetration to the bottom of limnic systems, triggering reductions in the numbers and masses of other, native, plants. Another was that they sucked up important nutrients such as nitrogen and phosphate compound, thereby reducing phytoplankton densities and, consequently, zooplankton [6]. As they died in the autumn and decomposed during the winter stagnation period, their increased oxygen consumption caused anoxic conditions leading to fish kills. During the period of most intensive growth in late spring, summer, and early fall, the *Elodea* also frequently consumed so much carbon dioxide that the pH of the water skyrocketed, killing off many pH-sensitive organisms [7].

Of course, the waterweed was also a potential new food source that some organisms were better suited to make use of than others. In much of central and northern Europe, the prime benefactor was the great pond snail (*Lymnaea stagnalis*), an air-breathing lung snail. Among fish, although in Europe there is no typical vegetarian species, the partially vegetarian common rudd (*Scardinius erythrophthalmus*) is the one that seems mostly to have benefitted from the new food source. Young pike and pike perch, too, often increase their numbers in watercourses containing dense mats of Canadian waterweed, which offer them both good hiding places and hunting grounds.

Control Attempts

A large number of techniques have been used to try to control the mass development of *Elodea* [8]. Mechanical removal, as in the Swedish Brogårdsdammen, did little to stem the exceptional growth. Fractioning, with the aim of destroying the plants, did, if anything, have the opposite effect and tended to increase growth and spread.

Herbicides would, indeed, kill the Canadian waterweed, together with most other aquatic plants, but once the herbicide was broken down or flushed out, the *Elodea* would be the first to return and, in the absence of competitors, would quickly dominate even more than before.

Biological control with grass carp (*Ctenopharyngodon idella*) worked reasonably well, but was only practiced from the late twentieth century in the relatively few ponds and lakes that then saw mass development of the Canadian waterweed [9]. Besides Poland and Russia, Denmark and Sweden are countries where this Chinese herbivore fish successfully controlled *Elodea* invasions.

Then, What Happened?

In the real majority of cases (some comments will be given below to the rare exceptions), within a decade or so after their introduction in any body of water, the exceptional growth of the Canadian waterweed subsided, leaving the plant a common feature of practically any freshwater body in Europe but no longer a dangerous, dominating species that suffocated others and changed ecosystem dynamics.

In Germany, the problem with the "wasser-pest" invasion was seen as being over in 1910, when Hermann Löns, a famous journalist, author, and naturalist, ironically recalled earlier doomsday sentiments in an article: "A dreadful howling and yowling rose all over the land, because the fearsome day seemed to be close at hand, when all freshwaters of Europe would be full of this weed, so that no ship could sail, no man could bathe, no duck could dive, and no fish could swim" [10].

The Canadian waterweed is a prime example of an organism that, after having enjoyed an initial period of uninhibited growth, was brought under control by native members of the ecosystem that happily turned to feed on the new plant. Most assessments in Europe today of *Elodea canadensis* as an invasive species conclude that it is no longer a significant problem and that it now forms a normal and integral part of most aquatic ecosystems.

The prime actor and beneficiary is the great pond snail, as mentioned above, and to some extent the common rudd. In addition, some aquatic nematodes attack the buds and can substantially delay the waterweed's otherwise explosive spring growth. Canadian waterweed has also become one of the favorite foods of the European beaver, just as it is for their Canadian cousins. And plant-eating aquatic birds such as swans and geese have added the Canadian waterweed as a favorite item on their menu. In many parts of Europe, these large birds, like beavers, have had a dramatic population rebound after earlier being heavily hunted. Among geese, another Canadian invasive species, the Canada goose (*Branta canadensis*), has established a strong European presence. For them, the Canadian waterweed is a familiar food source from back home.

Some biologists draw attention to the fact that in Europe, only female plants of the Canadian waterweed are present. Thus, even if the plants occasionally flower, only vegetative reproduction takes place, and the recombination of genes that sexual reproduction achieves does not take place. This makes the plant less capable of adjusting to changing environments; biologists see this reduced adaptability as a possible contributing factor to the relatively fast decline of *Elodea* in Europe.

Another factor behind the decline of *Elodea canadensis* is that Europe in the twentieth century also received another *Elodea* immigrant from North America, *E. nuttallii*, the Western or Nuttall's waterweed. It is a close relative of the Canadian one, and they are not easy to tell apart, which explains why it took a long time to detect its presence. The current trend is that *E. nuttallii* is taking over from *E. canadensis* in places such as in Britain [11], as well as in Lake Mälaren outside of Stockholm in Sweden. So far, it has not shown the same explosive growth that the Canadian waterweed initially did, but it frequently seems capable of forming denser masses than *E. canadensis* does today.

What About the Exceptions?

There is one type of place, generally small bodies of water, where the mass development of the Canadian waterweed seems to dominate for much longer time than the typical decade seen elsewhere. These are man-made, excavated ponds or restored eutrophic lakes in which sediments have been dredged out, creating a bottom of blue clay. The Kemnader See outside Bochum in Germany is a well-published example. The exact reasons for this are not known, but as such bodies of water are so small and so few, they do not matter much to the broader picture.

Economic Effects

Today, the economic aspects are totally insignificant. Even if sudden mass development of Canadian waterweed still occasionally occurs, during which the plant colonizes a new pond or lake, the big picture is that this once-alien and invasive species relatively quickly becomes an integrated member of the European aquatic ecosystem, leading to changes in abundance of some native species, but without serious negative effects.

At the height of its invasiveness, though, the situation was quite different. Although no Europe-wide assessment of damage and control costs seems to exist, local reports indicate that overall annual costs may have exceeded $100 million in today's currencies. As the plant spread eastward, this burden shifted from country to country.

References

1. Simpson, D. A. (1984). A short history of the introduction of Elodea in the British Isles. *Watsonia, 15*, 1–9.
2. Kelcey, J. G., & Müller, H. (2011). *Plants and habitats of European cities*. New York: Springer Science.

3. Eriksson, J. (1878). Amerikanska vattenpesten vid Skara. *Svenska Trädgårdsföreningens Tidskrift*.
4. Hylander, N. (1971). Prima loca plantarum vascularium Sueciae. Första litteraturuppgift för Sveriges vildväxande kärlväxter jämte uppgifter om första svenska fynd. Förvildade eller i senare tid inkomna växter. *Svensk Botanisk Tidskrift, 64*(Suppl), 1–332.
5. Bazarova, B. B., & Pronin, N. M. (2010). *Elodea Canadensis*, Michaux, in the watershed of the Arctic and Pacific oceans. *Russian Journal of Biological Invasions, 1*(4), 243–250.
6. Erhard, D., & Gross, E. M. (2006). Allelopathic activity of *Elodea canadensis* and *Elodea nuttallii* against epiphytes and phytoplankton. *Aquatic Botany, 85*, 203–211.
7. Roerslett, B., Berge, D., & Johansen, S. W. (1984). Mass invasion of *Elodea canadensis* in a mesotrophic, South Norwegian lake—Impact on water quality. *Proceedings of the Congres of International Association of Theoretical and Applied Limnology, 22*(2), 2920–2926.
8. Vernon, E., & Hamilton, H. (2011). *Literature review on methods of control and eradication of Canadian pondweed and Nuttall's pondweed in standing waters* (Scottish Natural Heritage Commissioned Report No. 433).
9. Zehnsdorf, A., Hussner, A., Eismann, F., Rönicke, H., & Melzer, A. (2015). Management options of invasive *Elodea nuttallii* and *Elodea Canadensis*. *Limnologica—Ecology and Management of Inland Waters, 51*, 110–117.
10. Löns, H. (1910). Hannoversche Tagesblatt 9 October 1910.
11. Simpson, D. A. (1990). Displacement of *Elodea Canadensis* Michx. By *Elodea nuttallii* (Planch.) St. John in the British Isles. *Watsonia, 18*, 173–177.

Himalayan Balsam in Europe

In the botanical history of the Himalayan balsam plant, an Englishman named John Forbes Royle played an important role. He was the station surgeon for the East India Company and superintendent of the Botanical Garden in Saharanpur, Uttar Pradesh, India, between 1823 and 1831. He had the plant, together with many others native to the area, collected and brought there for him to examine and propagate/cultivate [1]. Royle took a particular interest in the Himalayan balsam and was the first to describe it scientifically and name it. In 1837, he was appointed professor at King's College in London and left India for Great Britain, in all likelihood bringing seeds of *Impatiens glandulifera* with him. Although practically all sources give 1839 as the year in which the Himalayan balsam was first planted in Kew Gardens (the most important center of botanical research in Europe for several centuries) and thereby introduced to Europe, given the short storage limit for the seeds to remain viable, it seems more likely that the introduction actually took place a year or two earlier [2].

Whatever the exact date of its introduction, records show that the first naturalizations were reported from Hertfordshire and Middlesex in 1855 [3].

The Plant Itself

In its native habitat in the Western Himalaya (present-day India, Pakistan, and Nepal), the Himalayan balsam grows at high elevations (1800–4000 m above sea level) along watercourses and in sparse forests. The balsam is an annual plant that can reach a height of 2.5 m. Its hollow stem is thick, jointed, and juicy, normally with few branches, but it can be bush-like, especially as regrowth when the stem has been cut. A system of supporting roots gives the tall plant stability, and the large blossoms, which somewhat resemble orchids, sit in clusters and produce copious amounts of nectar that attracts numerous fertilizing insects, particularly bumblebees. Their color varies from dark pink or violet to white, and the species can self-fertilize. The fruit comprises so-called "crack capsules" that explode at the slightest touch

A. Jernelöv, *The Long-Term Fate of Invasive Species*,
DOI 10.1007/978-3-319-55396-2_4

when mature, ejecting the seeds up to five meters. Seed production is high, normally around 700–800 per plant, but cases have been recorded in which a single plant produced several 1000 seeds. The seeds are short-lived and generally either sprout the following spring or die, although there are reports of 18 months' survival time. The entire seed bank in an area generally germinates at the same time, and initial growth is vigorous—after some 100 days, the plants can be more than 2 m high [4].

Himalayan balsam (*Impatiens glandulifera*) (Photo from http://www.wildplantforager.com/)

The Himalayan balsam thrives in moist soil and can stand shade, but is sensitive to drought. It often grows in thick stands, with up to 50 plants per m^2. In its native habitat in the Himalayas, it is reported to be frost tolerant, but that trait does not exist among its immigrants to Europe, according to studies of the British population [4]. Corresponding studies in Germany also conclude that "in Europe *I. glandulifera* plants of all ages are frost-intolerant. Usually, all adult plants are killed by the first frost in the autumn and seedlings are killed by late frosts in the spring" [5]. If that is the case, frost intolerance would define a northern border for the invasion. However, given the current spread to areas north of the Arctic Circle in Norway, Finland, and Russia, this may nowadays not be the case for the whole of the European population. At the same time, for the seeds to germinate, they need a "cold-kick" of some 5 °C for several weeks, which could define a southern border of the invasion.

The natural spread of invasive plant species over longer distances is aided when seeds are carried by water, tumbling like small stones along the bottom of fast-running creeks or rivers during high water flow events, eventually settling on the bottom in calmer sections and finding themselves on riverbanks when the water recedes [6].

In the autumn, when the plants die off, thick layers of organic debris are left and can be an obstacle for the early growth of less sturdy plants the following spring.

A stand of Himalayan balsam (Photo from http://www.wildplantforager.com/)

The Further Spread in Europe and Current Distribution

The further spread within Europe was, like its initial introduction to the continent, to a large extent a human affair. Seeds were distributed from London to a number of botanical gardens, such as in Berlin and Moscow, and from there to other botanical gardens and seed handlers for further distribution to private gardeners. In Stockholm, the Himalayan balsam appeared already in 1842, in the garden of the Garden Association, the seed having come from Berlin. In the catalog of seeds for sale from the botanical gardens of Uppsala, it was first recorded in 1852, and over the period from 1885 to 1953, it was a standard item in the catalogs of private seed handlers [7]. The pattern was similar over much of Europe.

As it had done in England, the plant escaped the confinement of gardens almost everywhere and established "wild" populations, sometimes aided by advice to seed customers such as this from Sweden: "It is further recommended that it is planted along banks of water-courses, where it might succeed to self-sow" [8]. And self-sow it did indeed, spreading downstream with the water flow and sometimes even transcending national borders. Thus did a Swiss population, naturalized in 1904, colonize a part of Germany via the Rhine River, some time during World War I. By 1920, it was already common in some places in the southwest of the country.

Back in Britain, naturalized specimens were also soon found in other places, such as an area near Manchester in 1889, followed by, inter alia, Berkshire, Buckinghamshire, North Devon, and County Durham. Its rapid spread was undoubtedly promoted by descriptions such as "It's one of the many desirable hardy plants from the North of India, which have of late been introduced," and "Few annuals are better worthy of a place in every good-sized flower garden" [4]. Many owners of such gardens heeded the call, and the Himalayan balsam soon became a widely distributed garden plant on the British Isles. Another group that enthusiastically spread the plant was beekeepers, who found that the late-flowering, nectar-rich plant contributed both to honey production and to bee winter survival. From gardens, the plant spread and established its presence in most corners of the country, and from there, it went further, especially along watercourses, till its once-sporadic distribution had turned into a practically ubiquitous presence on the islands. There seems to have been something of a time lag between the first releases to the wild and the population explosion and spread, with 1933 being the point of inflection on a time axis [9].

In Sweden, the first out-of-garden occurrences were reported from Lund in 1873 and on the Stockholm island of Djurgården in 1917. During the 1920s, there were many reports of small stands or single specimens growing "where garden waste was dumped," which was often close to rivers. Around 1940, the Himalayan balsam was naturalized (a term not yet in use at the time) in many places around Uppsala, Stockholm, and Gothenburg, and sightings had been reported from many places in South and Central Sweden, including on the Baltic islands of Öland and Gotland [10]. From the 1960s through the 1990s, the spread intensified, and its extent came to encompass most of the southern third of the country as well as the Baltic coast up to the Finnish border. An important factor behind the acceleration is thought to have been changed agrarian practices along riverbanks: animal grazing was practiced less and less, and hay harvesting on the embankments had all but ceased. These had previously kept the Himalayan balsam somewhat at bay. Since the turn of the millennium, the spread has continued northward and into the hitherto unconquered areas in the flatlands and forests in the south, although its speed may have diminished somewhat.

In some places where the plant was abundant many decades ago, it is no longer present. Often, this is associated with changes in soil moisture. In other places, such as around the mouth of Suseån River on the Swedish west coast, the plant was first reported in 1928 and still grew profusely in 1998 according to a letter to the investigators [7]. **In 2015, however, when I visited the site together with professor Ulf Wahlgren, the Himalayan balsam was fully absent from the very mouth of the river, grow sparsely and low at a distance of 150-300 m from the sea in apparent tough competition with reed, sap and meadow-sweet. Upstream that and up to 2000 m from the sea, the Himalayan balsam grow in somewhat denser stands covering up to 30 m2each, together amounting to some 600 m2. The stands were all in the zone that was likely to be submerged during spring flood and also contained field bindweed (Convolvulus arvensis), nettles and wild raspberries. The density of balsams in the stands were 20-40/m2 and they constituted 30-70% of the plants there. Given the many signs of reoccurring high waters, it seems likely that the vegetation on the banks would at intervals be eroded a way and reseeded, forming a transient floral community**.

In Finland, the Himalayan balsam arrived at the Botanical Garden of Helsinki from that in Moscow at the end of the nineteenth century, but it took until 1947 before the first naturalized specimens were observed. Since then, it has spread and is now fairly common in the south and central parts of the country. It has also, at an accelerating pace during the last 20 years, moved north and is now found even above the Arctic Circle [11].

From the Czech Republic—and earlier from Czechoslovakia—the spread of the Himalayan balsam is similarly well documented [12]. The oldest record of an escape into the wild is from Northern Bohemia in 1896. Although the speed of dispersion increased after 1936, with an inflexion point that year similar to the British one in 1933, until the 1950s *I. glandulifera* occurred only at scattered points across the country. A close check of the earliest records of the Himalayan balsam in the Czech Republic showed that it was especially popular as a garden plant in the areas then populated by the German minority in the country. Many of the first reports of spontaneous occurrence came from disturbed sites in and around those settlements, after which the plant then spread downstream along nearby rivers [6].

Less use of riverbanks, as noted above, coupled with eutrophication of watercourses (especially with nitrogen) and increased deposition of organic sediments and debris on riverbanks are factors that contributed to the accelerating dispersion of the plant from the late 1930s. The increase was particularly strong in the 1960s, and by the mid-1990s, it was estimated that the Himalayan balsam could be found on 56% of the total length of Czech rivers [12]. Fifteen years later, the estimated proportion of rivers that were occupied had risen to 70%, and it was assumed that the expansion would continue [13]. It was also found that the rivers, investigated in detail, did not differ in the occurrence of the species along their banks, but there were some significant variations between the penetration of *I. glandulifera* further from their banks, both in terms of the lateral dimension in the main valley and upstream along the tributaries and the time since the first occurrence of the species in the vicinity of a given river.

Besides the countries already mentioned, the plant is also present as a naturalized species in the following European countries: Austria, Belgium, Bulgaria, Croatia, Denmark, Estonia, France, Germany, Hungary, Ireland, Latvia, Lichtenstein, Lithuania, Montenegro, the Netherlands, Norway, Poland, Romania, Russia, Serbia, Slovakia, Spain, Switzerland, and Ukraine (and in all likelihood also in Luxemburg and Slovenia, although regarding Greece, Italy, and Portugal, there is less certainty). In Norway, the range extends up to Tromsö at 69°,40′ N (300 km north of the Arctic Circle) [14]. In Russia, after explosive development and diffusion in the last few decades, *I. glandulifera* has become one of the most common alien species, found from the northern regions of Murmansk and Karelia in the north to the Caucasus in the south. It occurs on floodplains and along riverbanks, in wetlands and along roads, and in different human-made habitats [15]. In Northern Ossetia, in the Caucasus, it is found at high elevations (1100 m), much as in its native Himalaya [16].

Ecological Effects of the Himalayan Balsam

Environmental organizations, besides creating awareness of man's impact on global systems and collecting money for far-away projects, also engage volunteers in practical, collective work on local projects. This is often an important way to recruit young persons to the cause. Suitable projects for such activities are not always easy to find, and uprooting of Himalayan balsam has become one of the most common activities for such "grassroots" campaigns in many countries. To motivate people, information material, often called fact sheets, is distributed. On many "invasive species fact sheets," the ecological effects of the Himalayan balsam are summarized in words such as the following [17]:

- Creates dense stands that prevent native plants from establishing and reduces biodiversity and ecological value of land
- Prolific nectar producer, drawing pollinators away from surrounding native species
- An annual species that can aggressively replace native perennial plants along riverbanks, leading to soil erosion

Some aspects of the above description are undoubtedly true: Himalayan balsam creates dense stands; it is a prolific nectar producer; and it is an annual species. When it comes to the stated negative consequences, however, the entire matter becomes less clear. Does it really cause a loss of biodiversity and ecological value? Are pollinators drawn away to the detriment of native plant species? Does winter soil erosion increase when annuals die in the fall and before the next year's seedlings come up in the spring?

Checking the few references on such fact sheets leads, more often than not, to publications in which the authors support their statements by referring to someone else's paper, which in turn provides another reference, and then another, until in the end a paper is eventually found in which someone had only hypothesized that such negative effects could occur. Actual studies do exist, but they are surprisingly often not included in the reference loops. Furthermore, many of the existing studies have a potential built-in bias, as the method used compares sites populated by the invasive balsam with similar sites on which native plants are present. A long-term study of the situation before and during—and possibly after—the invasion would be more reliable, as "similar" sites might not be as similar as the researchers would like to believe, e.g., the reason why the Himalayan balsam did invade certain stretches of a floodplain or riverbank and not others could be preinvasion disturbances of those sites. If so, then the "lower diversity," "ecological value," and higher "soil erosion" might be a reason for, rather than a consequence of, the invasion.

Plant Species Diversity

When there is a real study at the end of the chain of references to studies on the impact of the Himalayan balsam on plant species diversity, it is often one from the northeast of England. There, a series of replicated *Impatiens glandulifera* removal experiments were undertaken in riparian habitats in order to address the impact of *I. glandulifera* on plant species' richness and evenness of distribution using local and habitat scales [18]. When the results are summarized for publication, they often appear as follows:

"River banks densely-colonized by Himalayan balsam have been shown to have reduced plant diversity by up to 25% (Hulme and Bremner, 2005)" [19].

The authors do, more or less, reach that conclusion, but they also state that "Non-native species accounted for more than one-third of the additional species found in removal plots. The non-native species that benefited most from *Impatiens* removal included the widespread perennial herbs *Lamium album, Myrrhis odorata, Petasites fragrans* and *Tanacetum parthenium*. Consequently, removal plots had a significantly higher proportion of non-native species than those where *Impatiens* was present." From their experiment, they also concluded that "although *Impatiens* reduces the native species diversity in open and frequently disturbed riparian vegetation, many of the species negatively influenced by *Impatiens* are widespread ruderal species. Furthermore, management may lead to compensatory increase in the abundance of other non-native species and thus fail to achieve desired conservation goals" [18].

In a similar study in the Czech Republic, a similar result was reached. There, the effects on riparian plant communities of *I. glandulifera* along invaded banks of six rivers were studied by removing the Himalayan balsam in some areas and comparing it to "un-invaded" communities and "uncleared" spots. Un-invaded spots harbored up to 23% more species than thickly invaded ones, but other effects were not apparent. Once *I. glandulifera* was removed, communities recovered without any lasting consequences for species diversity. It was concluded that "*I. glandulifera* exerts negligible effect on the plant characteristics of invaded riparian communities, and hence represents little threat to the plant diversity of invaded areas investigated during this study" [20].

In Germany, in the wetlands of the middle section of the Ruhr Valley, 80 permanent plots were investigated phytosociologically from 1989 to 2001. Data on neophytes were collected and used for an evaluation of the dynamics of spread at the landscape level. One of them was *Impatiens glandulifera*. Extreme fluctuations were recorded, with temporary colonization of reed communities in large quantities in some years and total regression in other years. Over the 13-year observation period, the Himalayan balsam occupied an increasing number of plots during the first 7 years, but this was followed by a pronounced decrease in the period from 1997 to 2001 [21]. With regard to species diversity in plots dominated by one species, the author found that *I. glandulifera* is no worse than competing native species, as illustrated in the below republished Table 1.

"In most cases," he states, "the indigenous species *Urtica dioica* or *Phalaris arundinacea* would become dominant in the absence of *Impatiens glandulifera*."

Table 1 Associated floristic diversity in dense stands of four species that tend to dominate communities in the Ruhr Valley, calculated from all vegetation plots with dense stands (i.e., cover value of the respective species 75–100%), showing specimen number per plot

Dominant species	Minimum	Mean	Maximum	Standard deviation	No. of plots
Impatiens glandulifera	5	9.8	15	3.7	8.0
Urtica dioica	6	12.9	24	5.5	21.0
Glyceria maxima	2	8.5	18	6.0	8.0
Phalaris arundinacea	1	5.5	9	2.8	6.0

The overall conclusion is that "apprehensions that *I. glandulifera* or other neophytes could outcompete native species in certain habitats can turn out to be a delusion caused by temporarily large quantities of fluctuating neophytes. There is no reason to expect special challenges for nature conservation resulting from neophytes in the Ruhr Valley" [21].

In a study in the Czech Republic, the effects of the invasion of 13 invasive neophytes (alien species introduced after anno 1500) in a variety of plant communities were examined with regard to species composition and diversity in the invaded communities. It concluded that in the riparian habitats that *Impatiens glandulifera* invades, where native communities are dominated by tall nitrophilous species, the impact of *I. glandulifera* is not significantly different from that of those native dominants. Thus, the Himalayan balsam was one of two invasive species that did not reduce species diversity, richness, or evenness in the invaded plots [22].

A somewhat special case of competition has been found between *I. glandulifera* and the two other *Impatiens* species present in many European countries, the likewise invasive *I. parviflora* and the native *I. noli-tangere*. They all occur in similar habitats and share basic life-history characteristics. When all three are present at a forested site, they each tend to dominate particular sections of it, with the presence of *I. noli-tangere* being strongly correlated with high soil moisture that of *I. parviflora* with high tree canopy cover and low soil moisture. *Impatiens glandulifera* exhibits a unimodal response to tree canopy cover, avoiding both very shaded and fully open sites. In a study in the Czech Republic, it was found that the current-year abundance of all species was negatively related to those of congeneric species and that the inter-year variation in dominance was considerable and mainly due to spring moisture fluctuations. The authors found the lack of a relationship between the occurrence of *I. glandulifera* and soil moisture rather surprising, as the species is commonly reported as requiring high soil moisture. A further spread of the invasive *Impatiens* species would then, anyhow, in all likelihood mean a narrowing of the niche for the native *I. noli-tangere* [23].

Effects of Himalayan Balsam on the Diversity of Plant-Associated Fauna

There are a number of studies of the nectar production of Himalayan balsam, with a more or less universal conclusion that it is high compared to almost all native plants. There are also a large number of studies of the nectar-seeking, pollinating insects that visit the plant, almost all of them finding a rich and diverse community of arthropod visitors. From there, however, there are differing interpretations of what that means. Conservation biologists such as those who produce the type of texts exemplified above tend to conclude that if the pollinators visit the invasive plant species, it will be at the expense of native species, with a resulting reduction of pollination success among those. There are hundreds of papers drawing that conclusion, but a review of the literature found no actual experiments or observations that verify a loss of native plant fertility for that reason. On the contrary, several studies found no such effect. In a study around the city of Göttingen in Central Germany, for instance, the authors concluded that their data provided no evidence that *I. glandulifera* outcompetes native plants for pollinators nor that it had any significant effects on their seed set [24].

An alternative view, much less often expressed but better supported by experiments and observations, is that the availability of a rich nectar source will support a richer and more diverse fauna of pollinators [25–27] also to the benefit of native species. Given that at least in the northern part of the range of invasion of the Himalayan balsam late bloomers are relatively few, the "competition-for-pollinator" argument may be weaker and that for the "value-of-support-of-a-richer-and-more-diverse-pollinator-fauna" stronger.

In a study of terrestrial gastropods in plots invaded by the Himalayan balsam, as compared with un-invaded ones in a deciduous forest in Switzerland, it was found that species richness and gastropod abundance were significantly higher in the invaded plots [28]. In a similar study of *Collembola* and *Acari* species in leaf litter and soil samples, *Collembola* species richness and abundance were not significantly different in invaded compared to un-invaded plots, while the abundance of leaf-litter-dwelling *Acari* was higher in balsam-invaded ones [29].

Partly in contrast to those results, in the UK during 2007 and 2008, the differences in the above- (foliage- and ground-dwelling) and belowground invertebrate communities in nine un-invaded plots and nine plots invaded by *I. glandulifera* were evaluated. In the foliage-dwelling community, all functional feeding groups were less abundant in the invaded plots, and the species richness of *Coleoptera* and *Heteroptera* was significantly reduced. In the ground-dwelling community, herbivores, detritivores, and predators were all significantly less abundant in the invaded plots. In contrast, these functional groups in the belowground community appeared to be largely unaffected and even positively associated with the presence of *I. glandulifera* [30].

A group of organisms that could be expected to benefit from the insect fauna attracted to stands of Himalayan balsam by the rich supply of nectar is spiders. Only one scientific study could be found that, with the focus on the effects of coal mine drainage, indirectly dealt with the effects of the plant on the diversity and richness of this group [31]. However, there are numerous reports on garden blogs and Facebook and Twitter posts of amateur observations of the presence of many and unusual spiders on balsam stands [32].

Himalayan Balsam on Riverbanks and Increased Erosion

Two, at least at first glance, contradictory claims are routinely made about the effects of stands of Himalayan balsam on riverbanks. One is that the thick layers of debris of the previous year's foliage prevent other plants from coming up in the spring. The other is that when the annual plant dies off in the fall, riverbanks are laid bare and exposed to erosion. The circle may be squared if one assumes that the claims relate to different stands of Himalayan balsam, wherein the first refers specifically to stands on non-flooded grounds and the second to such where water first sweeps away the previous year's annual vegetation spoils and then the soil itself. While there is hardly any observational or experimental support for the first claim, such a study supporting the increased erosion claim was recently published, although subject to the type of criticism of the methodology mentioned above.

Along the riparian zone of a small watercourse in northwest Switzerland, changes in the soil surface profile (SSP) were measured at six separate locations, each contaminated with *I. glandulifera*, while the same number of nearby locations supporting natural vegetation was also examined. The measures were repeated at all 12 locations on seven separate occasions, from October 2012 to May 2013. This covers the time before the winter dieback occurred, through to the germination and seasonal regrowth of new plants. Results showed erosion from balsam-contaminated sites was significantly greater than erosion from topographically comparable reference sites supporting natural vegetation [33].

Some Observations of the Spread and Impact of the Himalayan Balsam in the Flatlands of Southwestern Sweden

From my own experience, I can tell the story of the arrival of the Himalayan balsam to the small village of Järpås in the flatlands of Västergötland in southwestern Sweden. Sometime in the late 1960s, I first saw the plant at a designated dumping area for garden waste at a small wet meadow in a forest at the edge of the village. At the time, the plant was new to me and I was impressed by its height. I also observed how attracted bumblebees were to it. Five years later, the entire meadow, approximately 50 × 25 m, was covered by the new plant. After another 10 years, the situation was the same, with the exception that boys from the village and nearby farms had discovered the small explosions with which the plant expelled its seeds and made their way through the thick growth in the autumn just to trigger them. With some of the seeds possibly sticking to them when they left the area, the boys could have been agents of further spread of the balsam.

By the mid-1980s, the plant had appeared in some village gardens, appreciated and cared for by the owners. By the late 1980s, there was a small stand at the little creek that transverses the village. Another few years later, there were thick stands along the banks at every sun-exposed place that was not cut, for five kilometers downstream. By this time, sentiments had changed and most villagers were already doing their best to clear their gardens of the invader.

The areas that the Himalayan balsam took over had previously mostly had thick stands of meadowsweet (*Filipendula ulmaria*) or, on some stretches, nettles

(*Urtica dioica*) or the common reed (*Phragmites australis*). The previous dominants were not all gone, but constituted a minor part (10–20%) of the plants, with the new dominant making up some 60–80%. At one spot, a stand of giant hogweed (*Heracleum mantegazzianum*) had been cut down the autumn before the Himalayan balsam appeared on that stretch, thereby, likely, paving the way for the new invader.

In the summer and early fall of 2014, I inventoried a dozen blooming Himalayan balsam stands along the Järpås creek for affiliated fauna. Bumblebees, bees, butterflies, hoverflies, flies, and moths were among the insects looking for nectar, but the most common pollinating species was, surprisingly, a wasp (*Vespula germanica*). The wasps' appearance was whitish, as if sprayed with powdered sugar, betraying their collection of pollen. In addition, in every balsam stand, the white-lipped snail (*Cepaea hortensis*) was present in large numbers, mostly with several individuals on each plant, while the invasive Spanish forest slug (*Arion vulgaris*), otherwise a ubiquitous pest in the area, was virtually absent. In some stands, the black bean aphid (*Aphis fabae*) covered the very tops of the stems and leaves of many plants, obviously overcoming the plant's chemical defenses. I also noted that several pairs of wagtail (*Motacilla alba*) had eggs or young ones in their nests along the creek in August, suggesting a third litter for the year, although in Sweden these bird species normally only have two litters.

Pollen-covered wasp on a Himalayan balsam flower (Photo by Paul Iddon http://www.wildaboutbritain.co.uk/)

On the meadow by the garden waste dumping site on which I had initially observed the first thick stand of Himalayan balsam, the plant was still present in the summer of 2014, but wild raspberries (*Rubus idaeus*) now dominated. The meadow also appeared to be less wet than it had been earlier, although that might have been due to ongoing maintenance work being done on a nearby forest drainage system.

Attitudes, Classifications, and Extermination and Control Attempts

Despite being highly invasive and having colonized most of Great Britain, conservation groups were for a very long time at ease with the presence of *I. glandulifera*. It was not listed as an unwanted alien species until the 1990s [34], when this view changed and the Himalayan balsam was classified as one of the four invasive species causing the most problems [35].

This pattern of initial acceptance followed by a change of heart has been repeated in many of the countries to which the Himalayan balsam has been introduced. It is probably coupled with the fact that the plant was widely regarded as a "garden jewel" well before it became a successful invader. It may also have to do with "the spirit of the times." Up until the mid-1900s, new species of flora or fauna were generally seen as enrichments, as long as they were not overt agricultural pests and did not present immediately recognizable health hazards. Gradually, a more and more restrictive view gained ground, and today species that have spread with human assistance are seen as unwanted by the orthodox, regardless of their impact on their new environments.

In Sweden, *I. glandulifera* is counted as one of the five most aggressive invasive plant species [11], and in Switzerland and Austria, it is also nowadays seen as a "problem species."

Following this change of heart, a large number of eradication or control campaigns have been undertaken targeting the Himalayan balsam. Generally speaking, the strategy has been to use the fact that the plant is an annual and the seeds short-lived, to uproot or cut the plants just above ground before they set seeds. In theory, that should, if done carefully, eliminate the plant from the locale in question. However, what may appear easy in theory and implementable on small plots such as gardens has turned out to be very difficult in nature, for example, on riverbanks. There, a single plant left uncut may reseed an entire area, or plants cut some centimeters above ground may survive, branch out, flower, and set seeds. Seeds may also come with water from unknown stands upstream. In addition, repeated physical removal is labor intensive and costly if done by paid labor. Overall, of a large number of attempts, few have been lasting successes, and in some cases, when the offending invasive plant has been removed, it has been replaced by unwanted, but native, species such as nettles.

Chemical control with herbicides is cheaper but much riskier, as other plants around and in the stands are likely to be effected too. Herbicides applied for Himalayan balsam control include glyphosate, 2,4-D amine, triclopyr, and picloram. In some cases, when balsam stands along rivers have been sprayed, the herbicides have been carried downstream and damaged aquatic plants kilometers away. As in the case of physical removal, reseeding of Himalayan balsam or its replacement by other unwanted alien or native plants after local chemical eradication may occur.

Generally speaking, no foolproof methods have been found thus far for large-scale eradication or control of the Himalayan balsam. Some hope for a biological control agent which is now being placed on a rust fungus that CABI—a not-for-profit British research organization—received permission to release on plots in Berkshire, Cornwall, and Middlesex in the summer of 2014, after 8 years of laboratory trials [36]. In 2015, the trials were extended to 25 new sites.

Not everyone sees the plant as a pest, though. Groups interested in "food from nature" have emerged as new supporters, together with the ever-more squeezed beekeepers, whose views can be expressed this way: "Thankfully, Himalayan/Indian balsam is here to stay. What a fantastic pioneer plant we have on our hands. The fact of the matter is that it is very well adapted to our climate, is edible, and it grows only where the ecosystem has been disturbed by human influence" [37].

At the country level, as well, opinions are not universally negative. In the Czech Republic, where numerous studies of the Himalayan balsam have been made, the nature protection problems associated with it are seen as small. In Germany, the comments are mostly more guarded: when the plant's invasiveness is described, it is often added that "the eye-catching dominant stands of Himalayan balsam develop only in late summer, so that other plants can grow and even flower and produce seeds, before they are shaded out." With special reference to nectar-seeking insects and their subsequent steps in their food web, it is even said that "for the fauna, the effects of *I. glandulifera* are mostly positive" [38].

Economic Aspects of Himalayan Balsam

There is hardly any direct economic damage attributable to the invasion of the Himalayan balsam. The sites it occupies are generally not of economic importance and neither are the plants it replaces. However, since it has been classified as an unwanted species and has attained pest status in many countries, attempts are being made to eradicate or control it, which come with a cost. No Europe-wide figures for the costs of eradication seem to have been calculated. But even though the manual labor entailed by eradication is mostly done by volunteers, the total annual cost of attempts being made by individual countries in all likelihood exceeds US$ 50 million for the continent as a whole.

A small economic benefit does exist in the form of an increased production of honey, guesstimated to amount to a few million dollars annually. But if, as most beekeepers and some scientists believe, the Himalayan balsam in general helps to sustain a larger fauna of pollinators, this would benefit a host of other plants, including agriculturally important ones that blossom in the spring and early summer before the balsam itself starts to bloom. The resulting economic benefit could then easily be an order of magnitude larger.

Current Trends and the Likely Future of the Himalayan Balsam Invasion in Europe

In some European countries such as the UK and the Czech Republic, the Himalayan balsam has already taken over most of the suitable riverbanks. In others, such as Sweden and France, the coverage is still much smaller. Currently, the riverside spread continues practically everywhere and not at a slow pace. There is nothing strongly indicating that this trend would soon be broken, so one has to postulate that, sooner or later, suitable habitats along more or less all rivers will be invaded. Once a specific site has been colonized, however, it need not remain so for long, as the dynamic of these habitats may make for rapid shifts in dominating species, as the Ruhr River case showed. In other cases, such as the Swedish Suseån, Himalayan balsam stands can seem to be permanent features, even on a timescale of almost a century.

In other habitats suitable for the Himalayan balsam, such as in sparsely vegetated swamp forests where no water flow will assist the spread, the advancement of the

Roe deer eating Himalayan balsam (Photo from [39])

plant is slow but steady, measured only in meters per year. Here, the stands tend to be less dense and more mixed with other plant species, both native and alien. A further slow but long-lasting spread is expected also in these habitats.

What, then, could turn the tide? Just at the horizon are a few candidates. The rust fungus mentioned earlier as a possible biological control agent is one. Others, plausible but possibly more likely to be long shots, are the black bean aphid seen blighting Himalayan balsam plants in Sweden and the diets of roe deer and elk (moose), which have been seen munching on Himalayan balsam in the UK and Finland, respectively [39].

References

1. Retrieved December 15, 2014, from http://www.icfre.org/pdf_files/7%20Herbaria.pdf
2. Retrieved December 17, 2014, from http://www.fera.defra.gov.uk/plants/documents/pucciniaPRAAppendix4.pdf
3. Retrieved December 16, 2014, from http://www.nonnativespecies.org/factsheet/downloadFactsheet.cfm?speciesId=1810
4. Beerling, D. J., & Perrins, J. M. (1993). *Impatiens glandulifera* Royle (*Impatiens roylei* Walp.) *Journal of Ecology, 81*, 367–382.
5. Sebald, O., Seybold, S., Philippi, G., & Wörz, A. (1998). *Die Farn- und Blütenpflanzen Baden-Württembergs*. Stuttgart: Ulmer.
6. Lhotska, M., & Kopecky, K. (1966). Zur Verbreitungsbiologie und Phytozönologie von Impatiens Glandulifera Royle an den Flusssystemen der Svitava, Svratka und oberen Odra. *Preslia, 38*, 376–385.
7. Larsson, C., & Martinsson, K. (1998). Jättebalsamin Impatiens Glandulifera i Sverige—Invasionsart eller trädgårdsflykting? *Svensk Botanisk Tidskrift, 92*, 329–345.
8. Pihl, A. (1884). Om balsaminerna. *Svenska Trädgårdsföreningens Årsskrift, 1884*, 161–164.
9. Perrins, J., Fitter, A., & Williamson, M. (1993). Population biology and rates of invasion of three introduced impatiens species in the British Isles. *Journal of Biogeography, 20*, 33–44.
10. Witte, H. (1941). Impatiens Glandulifera Lindeley såsom adventivväxt i Sverige. *Svensk Botanisk Tidskrift, 35*, 211–218.
11. Helmisaari, H. (2010). *NOBANIS—Invasive alien species fact sheet—Impatiens Glandulifera*. From: Online Database of the European Network on Invasive Alien Species—NOBANIS. Retrieved December 17, 2014, from www.nobanis.org
12. Pysek, P., & Prach, K. (1995). Invasion dynamics of Impatiens Glandulifera—A century of spreading reconstructed. *Biological Conservation, 74*(1), 41–48.
13. Malíková, L., & Prach, K. (2010). Spread of alien *Impatiens glandulifera* along rivers invaded at different times. *Ecohydrology & Hydrobiology, 10*(1), 81–85.
14. Alm, T. (2002). Kjempespringfrø (Impatiens Glandulifera)—En fremmed art i rask spredning i Nord-Norge. *Polarflokken 26*(2), 123–131. Tromsø.
15. Марков, М. В., Уланова, Н. Г., & Чубатова, Н. В. (1997). Род Недотрога.—В сб.: Биологическая флора Московской области, вып. 13. М., Изд-во МГУ. 128–168.
16. Комжа, А. Л., & Попов, К. П. (1990). Новые данные об адвентивной флоре Северной Осетии. *Ботанический журнал, 75*(1), 108–110.
17. Retrieved December 24, 2014, from http://www.invadingspecies.com/invaders/plants-terrestrial/himalayan-balsam/
18. Hulme, P. E., & Bremner, E. T. (2005). Assessing the impact of Impatiens Glandulifera on riparian habitats: Partitioning diversity components following species removal. *Journal of Applied Ecology, 43*, 43–50.

19. Retrieved from https://cabiinvasives.wordpress.com/2010/11/10/our-river-systems-are-undoudtly-one-of-the-most-diverse-habitats-found-within-the-british-isles-thier-potential-for-harbouri/
20. Hejda, M., & Pyšek, P. (2006). What is the impact of *Impatiens glandulifera* on species diversity of invaded riparian vegetation? *Biological Conservation, 132*, 143–152.
21. Kasperek, G. (2004). Fluctuations in numbers of neophytes, especially Impatiens Glandulifera, in permanent plots in a west German floodplain during 13 years. *Neobiota, 3*, 27–37.
22. Hejda, M., Pysek, P., & Jarosik, V. (2009). Impact of invasive plants on the species richness, diversity and composition of invaded communities. *Journal of Ecology, 97*(3), 393–403.
23. Cuda, J., Skalova, H., Janovsky, Z., & Pysek, P. (2014). Habitat requirements, short-term population dynamics and coexistence of native and invasive impatiens species: A field study. *Biological Invasions, 16*, 177–190.
24. Bartomeus, I., Vila, M., & Steffan-Dewenter, I. (2010). Combined effects of Impatiens Glandulifera invasion and landscape structure on native plant pollination. *Journal of Ecology, 98*, 440–450.
25. Showler, K. (1989). The Himalayan balsam in Britain—An undervalued source of nectar. *Bee World, 70*, 130–131.
26. Stary, P., & Tkalcu, B. (1998). Bumble-bees (Hym. Bombidae) associated with the expansive touch-me-not, *Impatiens glandulifera* in wetland biocorridors. *Anzeiger fur Schadlingskunde Pflanzenschutz Umweltschutz, 71*, 85–87.
27. Nienhuis, C. M., Dietzsch, A. C., & Stout, C. J. (2009). The impacts of an invasive alien plant and its removal on native bees. *Apidologie, 40*, 450–463.
28. Ruckli, R., Rusterholz, H. P., & Baur, B. (2013). Invasion of Impatiens Glandulifera affects terrestrial gastropodes by altering microclimate. *Oecologia, 47*, 16–23.
29. Rusterholz, H.-P., Salamon, J.-A., Ruckli, R., & Baur, B. (2014). Effects of the annual invasive plant *Impatiens glandulifera* on the Collembola and Acari communities in a deciduous forest. *Pedobiologia, 57*(4–6), 285–291.
30. Tanner, R. A., Varia, S., Eschen, R., Wood, S., Murphy, S. T., & Gange, A. C. (2013). Impacts of an invasive non-native annual weed, *Impatiens glandulifera*, on above- and below-ground invertebrate communities in the United Kingdom. *PLoS One, 8*(6), e67271.
31. Paetzold, A., Smith, M., Warren, P. H., & Maltby, L. (2011). Environmental impact propagated by cross-system subsidy: Chronic stream pollution controls riparian spider populations. *Ecology, 92*, 1711–1716.
32. Retrieved from http://utgangspunktnykarleby.blogspot.co.at/2013/06/gurkspindel.html#comment-form
33. Greenwood, P., & Kuhn, N. J. (2014). Does the invasive plant, Impatiens *glandulifera*, promote soil erosion along the riparian zone? An investigation on a small watercourse in Northwest Switzerland. *Journal of Soils and Sediments, 14*(3), 637–650.
34. Usher, M. B. (1986). Invasibility and wildlife conservation: Invasive species on nature reserves. *Philosophical Transactions of the Royal Society of London B, 314*, 695–710.
35. Pearman, D. A., Preston, C. D., & Dines, T. D. (2002). *The new atlas of British and Irish Flora*. London: Botanical Society of the British Isles.
36. Retrieved December 30, 2014, from http://www.cabi.org/news-and-media/2014/cabi-releases-rust-fungus-to-control-invasive-weed-himalayan-balsam/
37. Retrieved December 31, 2014, from http://www.wildplantforager.com/blog/edible-weed-how-to-eat-himalayan-balsam-flower-and-use-the-stem-as-a-straw
38. Retrieved December 31, 2014, from http://www.neobiota.de/12639.html
39. Retrieved December 31, 2014, from https://www.flickr.com/photos/9578475@N02/9524045521/in/photostream/

The House Sparrow in North America

"Dear sister, it's a long time since I last wrote and I haven't heard from you either for years, but yesterday something happened that suddenly brought back so many memories from our childhood that I felt compelled to write. Don't worry, it was nothing bad. In fact, I can barely remember the day I felt so happy. Why? I saw a small bird from back home. You may laugh because in Småland those sparrows are so common that you hardly notice them, but here we haven't had any at all—till now that is. I hadn't really thought about that they were missing, but when I saw it, my heart filled with joy and my eyes with tears and I thanked God with all my heart." (From a letter from a Swedish emigrant to America to her sister in Sweden)

A Little About the Bird Itself

The house sparrow, *Passer domesticus*, sometimes in North America called the English sparrow, is a small bird in the family Passeridea that originally had its home in Europe, a large part of Asia, and Mediterranean North Africa. It's typically 16 cm long and weighs between 25 and 40 g. Young birds and adult females are bleak brown gray in color, while the adult male has more distinct brown, black, white colors. Different populations, sometimes seen as subspecies, differ somewhat in coloration and body size and form, from thick, round, and short legged to more slender and long legged. During the mating season, the color is a little brighter, and the male beak is black instead of dark brown.

A. Jernelöv, *The Long-Term Fate of Invasive Species*,
DOI 10.1007/978-3-319-55396-2_5

Male house sparrow (From Wikipedia)

The house sparrow generally lives close to humans and supposedly has done so since man started with agriculture and husbandry. It's fairly opportunistic when it comes to food, but seeds from cultivated or wild plants are usually the main feed for most of the year. When the young ones are newly hatched, the house sparrow turns to an insect diet for a while, if there is any chance at all to do so. When forced to accept a different diet, mortality among the young ones increases substantially. Dependent on climate, the house sparrow can have up to 4 L a year with 5–8 youths in each. The color of the eggs is white, light green, or gray with brown spots, which sometimes may cover most of the egg.

House sparrow eggs (From Wikipedia)

Incubation time is just short of 2 weeks, and the young ones leave the nest after just over another fortnight.

Natural enemies of the house sparrow are most predatory birds, cats, and weasels. Rats, crows, and squirrels might take eggs and hatchlings.

Female house sparrow with a chick fresh out of the nest (From 10000birds.com)

Through humans, the house sparrow has been introduced to a large part of the world where it originally didn't occur and mostly it has done well in the new areas. It's now considered to be the wild bird with the largest area of distribution.

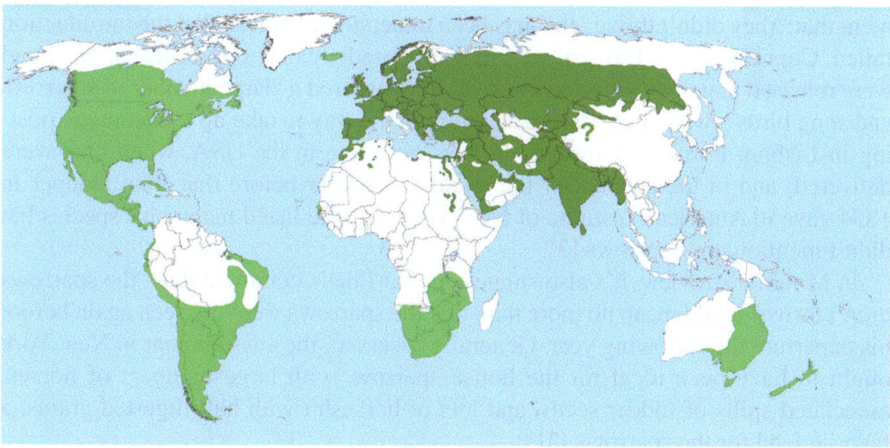

The global distribution of the house sparrow. Dark green areas are those of natural occurrence; light green ones were it was introduced by humans (From Wikipedia)

Introduction to North America

The most commonly told story about the introduction of the house sparrow to North America goes as follows: In the year 1850, the Brooklyn Institute brought in eight pairs of sparrows from England, kept them in a large cage over winter, and released them the following spring in Brooklyn, New York. The introduction didn't succeed, so the year after, another 100 sparrows were taken from England. Fifty of those were released on arrival in Narrows, while the others were caged over winter and let free on the spring of 1853 at the Greenwood Cemetery. These two later introductions, or at least one of them, succeeded and the species got established. The bird then spread by itself in circles from the release place, reaching a radius of 40 km after the first 5 years, 80 km after 10, and at least 150 km in 15 years [1]. Another tenths of introductions were made with house sparrows from Europe at different places in North America. Most of the birds came from the British Isles, but some also from, e.g., Germany. The introductions were partly made for nostalgic reasons, the immigrants wanted to see their familiar bird at the new homes, and partly in the hope that the sparrows would eat unwanted insects, in particular the so-called drop worms, larvae of the moth *Ennomos subsignaria.*

This often told story of the introduction, however, has been put in question [2], both with regard to the claim that the first attempt with the eight pairs in 1851 failed and with regard to whether the two later attempts, which are generally said to have been the successful ones, took place at all. At a closer look, both claims seem to stand on shaky grounds. It's all based on comments made by a man, Nicholas Pike, 35 years later. He certainly was a centrally placed person, when the first introduction took place, being the head of the Brooklyn Institute at the time. It's also his statement that "they didn't thrive" that has been understood to mean that the introduction failed. Concerning the 100 sparrows that supposedly came to the USA in 1852 and were released that and the following year, Pike ordered a "large number of sparrows and song birds", when he passed Liverpool on his way to take up a diplomatic posting in Lisbon, Portugal. However, he was not there in the USA, when they were delivered, and in the account of the releases the year before that John Hooper in 1854 gave to American Institute of New York City, he listed many bird species but didn't mention any sparrows [3].

In Moulton's review, it's also suggested that Pike's comment that "the sparrows didn't thrive" could mean no more than that the sparrows were not seen again before his departure the following year. Generally assessed, the environment in New York ought to have been ideal for the house sparrow, with large numbers of horses, associated spills of fodder seeds, and lots of horseshit with half-digested grains, a favorite feed for the sparrows [2].

It might thus just be so that the first introduction was successful and that the later ones, to which the success is generally ascribed, never took place.

In total, there were more than a 100 documented releases of house sparrows from European and newly established American sparrow populations up till 1886. Besides in various places on the US East Coast, where most of them took place, successful introductions were also made in Galveston, Texas (1867), San Francisco (1871 or 1872), and Salt Lake City (1873 or 1874). It's worth noting that the house sparrows in California, while successfully establishing themselves in San Francisco, didn't quickly spread much beyond it.

Also to Canada, the sparrows were introduced, e.g., in Québec (1854) and Ontario (1870).

After the first wave of introductions, which lasted for approximately 35 years, few additional ones are documented. By then, 1886, the house sparrow was well established east of Mississippi, and the birds continued to spread like rings on water, west, north, and south. During the 15 years, between 1871 and 1886, the area they populated in North America grows from a few 1000–1,340,000 km^2.

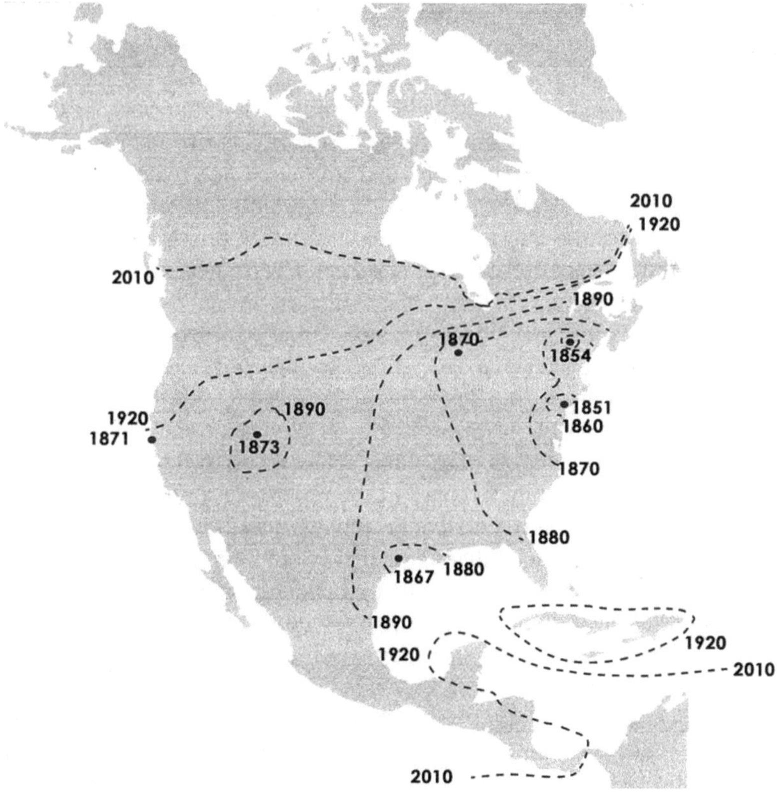

The spread of house sparrows in North America

At that time, a very intensive debate, which started around Boston and New York, had gone on for several years and spread to most areas the sparrows had conquered. It was referred to as "the sparrow war" and was about whether the damage the house sparrow did outweighed the good deeds and the pleasure of having it around the house—that is, if the birds should be classified as pests and bounties paid for killing them [4].

The house sparrow had initially a strong support among homeowners and in the outdoor-going sporty public, but in the other corner of the ring were two closely associated organizations, one of them a federal authority—the Department of Agriculture. The other was AOU, American Ornithologist's Union, an organization with the goal of studying and protect birds, in which employees of the department held leading posts. In the name of "biosecurity," they jointly went to the offensive against the "English" sparrow. Between themselves, they distributed the tasks so that the department talked about damage to agriculture, while AOU argued that the invading sparrows threatened native birds with extinction. For the purpose of winning the "sparrow-friendly" public for their position, the two organizations didn't shy away from using an immigrant-hostile arguments and play on the general anti-English sentiments and the immigration fatigue that prevailed on the East Coast. The debate was particularly rancorous, and the combatants excelled in demonizing their opponents. Journals like *Forest and Stream*, the leading outdoor magazine at the time, as well as daily quality papers like the *New York Times* and *Boston Globe* were awash with letters to the editor on the subject and also published a surprising numbers of featured articles and editorials. As a reaction to the pressure from their readers, they mostly took a "sparrow-friendly" position.

The task to resolve the infected question went to the ornithologist Hart Merriam and his assistant Walter Bradford Barrows at the newly formed Division for Economic Ornithology and Mammalogy at the US Department of Agriculture. (It was to become the latter whose name became most associated with the subsequent report.)

Their approach to the job was to design a questionnaire, print it in over 5000 copies, and distribute it widely to mostly farmers but also to sheriffs, journalists, and ornithologists. They asked for information on the occurrence of sparrows, their numbers, their behavior, and feeding habits. They received 3300 replies and these, in processed forms, constitute the base of the report. Of all the answers that commented on the question about economic plusses and minuses of the introduced birds, a few percentage thought that the sparrows did more good than bad for crops; a larger but still small percentage said that it ended up with plus-minus zero, while over 90% reported overall damage, sometimes substantial, to desired plants. Also with regard to the effect of the house sparrow on native bird species, almost all the reports were negative.

Those who defended the sparrows with one of the original arguments for introducing them, their consumption of insect pests dealt a devastating blow by the results of an investigation of the gut content of 522 house sparrows collected from localities covering a large part of the distribution area of the birds. Less than a 100 of them contained any insect remains whatsoever, and of those that did, two-thirds of them were from useful species or such without economic relevance.

Barrows report [5] was, when it came, seen by the experts as being of high scientific standing and fulfilling the highest demands on objectivity. Maybe, this was because so many of "the experts" were members of AOU and that Barrows gave full support to the views of the organization. Later-days experts have been less enthusiastic. Wing, e.g., wrote "Surprisingly enough, it's difficult to understand Barrows interpretation of his own data and what he writes doesn't explain his conclusions" [6].

Anyhow, it ended "the sparrow war."

Or so it seemed for 123 years. Interesting enough, a contribution to the debate about the benefits and damage caused by the house sparrow came in an article about what modern ornithologists would think about when they hear the words "sparrow war": Mao's campaign in 1958 against "the flying rats," the house sparrows, of which there were several 100 million, if not a billion, in China at the time. The idea behind the whole thing was to eliminate the birds as competitors to humans for grain, and the main method to rid China of them was to constantly scare them and force them to fly till they fell down dead of exhaustion. For biologists outside China, it sounded like a joke, as it appeared totally unattainable to eliminate a mass-occurring bird species in a short time this way.

The campaign, however, was carried out with a huge mobilization of the Chinese masses, who blew trumpets, banged tin cans, yelled, and made noise in any way possible for weeks. This was combined with more conventional methods of bird hunting like shooting and netting.

According to official Chinese sources, this campaign, like everything Mao initiated, was described as a huge success. A billion sparrows fell from the sky and could be collected and consumed by the masses, who afterward in addition didn't have to see the sparrows eat the grain they laboriously grew. No ornithologist I knew, and not even some who were devoted Maoists in Europe, bought this story.

The more surprising was to read in the American *Smithsonian* magazine in 2012 an article by Rob Dunn [7], where the author fully accepted the communist-Chinese version of the "sparrow war" but saw it only as the first battle. In the second—and decisive—battle, the sparrows won, according to Dunn. Even if house sparrows mostly eat grain and other vegetable matter, there are times of the year, when the young ones are newly hatched, when they feed their offspring insects and eat it themselves. In Dunn's version of Mao's war on the sparrows, the victory in the first battle, the almost elimination of them, became the beginning of the defeat. Without insect-eating sparrows, even if for only a short time of the year, the insects got the upper hand and multiplied into previously unseen numbers, destroying harvests, causing the starvation of 35 million people, and contributing to the failure of "the great leap forward," according to the author.

Mao realized his mistake, Dunn wrote, and gave the sparrows legal protection in 1960. The article ends with the reflection that it's only when a species is gone that we understand the true value of it, when it's common we only see the drawbacks.

With this article, Dunn reopens the discussion, if the bad thing the house sparrow does to man is outweighed by its good deeds—the debate that Barrows appeared to have ended 123 years ago.

When Barrows' report came, the house sparrow was protected in most of the states where it occurred, while in two of them, there was bounty money to collect for killing them. The legal protections were quickly lifted, but bounties were hardly introduced (and abolished where it existed) as Barrows himself argued hard against them, mostly on grounds of the associated high administrative costs. Instead, he recommended voluntary "sparrow clubs," the task of which it would be to make life hard for the sparrows in every conceivable way, e.g., by systematic destruction of nests and nesting sites. Such organizations were also formed, and some operated on their mandate for a while, but enthusiasm faded when the activities didn't visually effect the continued spread of the birds.

The areal expansion of the house sparrow territory went on, but in the places first colonized, like the surroundings of Boston and New York, the populations had started to thin out, when Barrows' report appeared. Many observers in the core area of the sparrows in eastern USA estimated that the numbers peaked around 1890 and decreased thereafter [1].

In other parts of the country, e.g., New Mexico, to which the house sparrow arrived much later, the peak numbers were reached in the 1920s, and the decrease in numbers happened at the time, when horses were already gradually replaced by cars and motorized agricultural vehicles and machinery [8].

The Economic and Ecological Effects of the House Sparrows

In Barrows' report from 1889, the types of damages that the sparrows caused on grains, vegetables, and fruits, according to the 3300 returned forms, were listed. The reporters found these damages to be extensive, but no attempt was made to quantify them in economic terms. More than a century later, Pimentel [9] wrote about starlings and house sparrows as agricultural pests in the USA saying that "together they cause damages to an estimated value of $1 billion annually." In a table in another article of the same lead author [10] one can find that the starlings are considered responsible for damages in the range of $800 million, which leaves $200 million for the sparrows. To eat grain and destroy plants by consuming shoots and flowers are not the only type of damage that house sparrows cause the agrobusiness. By littering grain storage and transportation facilities like silos and train wagons with feces and feathers, they cause economic damage of the same magnitude according to some scholars [11].

Added to those figures should be the damage from the spread of infectious diseases attributed to the sparrows. According to one summary, these birds can be coupled to the spread of 29 different human and animal illnesses [12]. In connection with outdoor rearing of chicken, turkey, and geese, the house sparrow can cause problems in all three categories. When present in great numbers, they consume the feed intended for the domesticated birds, they litter, and they may spread, e.g., bird flu and bird tuberculoses.

The ecological damage is mostly in the form of house sparrows taking over from and squeezing out native bird species. Barrows listed 70 species that his reporters saw as suffering from the competition from the sparrows. The interactions between house sparrows and some of these species, like the American cliff swallow (*Petrochelidon pyrrhonota*), tree swallow (*Tachycineta bicolor*), and eastern bluebird (*Sialia sialis*), have later been specifically studied [13].

The most important point of conflict and competition is in regard to nesting sites, as their feeding habits only overlap marginally. The house sparrows have the advantage of nesting early and not being migratory. Thereby, they have time to occupy sites and build their nests before competing species arrive. Sometimes, they can also aggressively take over nesting sites, and there are reports stating that they evicted earlier inhabitants by pushing eggs and chicks out of the nest. The large-scale relevance of such observations is uncertain, as house sparrows almost entirely nest close to humans and most other species do not, and there are huge sparsely populated areas in North America.

Then What Happened?

At the onset of the twentieth century, the house sparrow had gone from being a desired bird to being an object of extermination campaigns and termed things like "flying rats." The territorial expansion continued, but the density of birds, the number per surface area, decreased in the core areas in Eastern USA. When the number of horses from the 1920s and onward dropped, thereby draining up the prime source of feed for the sparrows, their numbers fell even more.

In a study of the population developments for a number of bird species from late nineteenth century to the end of World War II in and around the towns of Toledo, Bowling Green, and Sandusky in Ohio, with bearing on a much larger area in North Central USA including North Illinois, South Michigan, Western New York, and Pennsylvania, the author, Moseley, reports the rise and fall of the house sparrows [14]. He notes that in his youth, the aggressive house sparrow was not present. Their arrival was preceded by a dramatic increase in the number of cliff swallows (*Petrochelidon pyrrhonota*), when the farmers started to build modern barns, which provided excellent nesting sites. When the sparrows arrived (in the late 1880s), they expropriated the nesting sites. "The swallows disappeared and were forgotten," he wrote. Also, other swallows like the purple martin (*Progne subis*) were hard hit by the competition, and their numbers dropped significantly.

Predatory birds, like the eastern screech owl (*Megascops asio*) and American sparrow hawk (it's unclear which species he refers to as the popular name is used for several species of the genus *Accipiter* as well as for the falcon *Falco sparverius*), however, strongly increased in numbers, as the house sparrows dramatically added to their feed base. This predation, according to Moseley, halted the explosive increase in the number of sparrows, and when a series of unfavorable factors came together in the following decades, their numbers dropped significantly. The continued decrease in the number of horses was obviously one such factor, but, in his view, not the most important. Instead, he highlights the introduction of the harvesters and the changing methods of agriculture, plus competition from another invasive species, the starling. They came to the area in the 1920s, and the numbers grow and grow till they stabilized in the 1940s. By then, the house sparrows were already uncommon.

In 1943, Wing published an estimate of the number of house sparrows in the USA giving the figure 150 million [6]. The report has been widely quoted and sometimes misused as the figure erroneously has been used both as an estimate of the peak number and the current one. In his analysis of the spread of the house sparrow, Wing arrived at the conclusion that the species 40 years after introduction, in the early 1890s, already covered the range it had another 50 years later, that is, more or less all of the USA, North Mexico, and Southern Canada. He also states that of all the breeding birds in North America, the sparrows only make up 3% and express a firm conviction that future studies will yield even lower estimates. In addition, he refers with irony to (not named) colleagues who talk about "billions of flying rats."

The many concurring reports of falling numbers of house sparrows from the late nineteenth or early twentieth century, some of which have been referred to above, make it reasonable to assume that the peak number was at least twice and may be three times higher than the 150 million that Wing estimated in the early 1940s. If one adds the house sparrows in Mexico and Canada, it's not inconceivable that the number of house sparrows in North America, when it peaked, touched half a billion.

During the second part of the twentieth century, the shrinking population of house sparrows was faced with another problem: competition from yet another invasive species but a native and much liked one. In the nineteenth and the first part of the twentieth century, house finches (*Haemorhous mexicanus*) were popular cage birds (and still are in Mexico). They were sold under the trade name "Hollywood finches," not an all-together unsuitable name as the natural area of distribution was the North American West Coast from Oregon to South Mexico. When the Migratory Bird Treaty Act was signed into law in 1918, it formally became illegal to keep the finches as cage birds, but as no one acted to implement the law, the trade and practice continued. Finally, in 1939, the New York authorities decided to stop the trade in the birds and to fine the owners of zoo shops that had them for sale. To avoid the fines, many owners simply let the birds out before the inspectors arrived. The released birds succeeded to survive and established a small population that during the 1940s and 1950s barely survived in public parks and private gardens in the vicinity of the release points. In the 1960s, something happened, and the number of house finches started to increase, and they spread all over New York City and beyond. At the end of the 1980s, they were all over the East Coast from South Canada to Florida [15].

As the western house finch population continued its explosive expansion, an observant bird watcher in Maryland in 1994 noticed that some finches at his bird feeder had swollen eyes. A couple of days later, he found them dead on the ground and sent them for veterinary tests. It turned out that they were infected with *Mycoplasma gallisepticum* a well-known and common pathogen on chicken that had not previously been found on songbirds. From 1994 to 1996, the disease spread to the whole eastern house finch population, and in total some 100 million finches succumbed, half of finches in the east at time. Since then, the house finches have acquired some degree of immunity against the disease, which has faded, and the eastern population has recovered and retaken its westward spread.

In the meantime, also the original western population had expanded northward to Washington and British Colombia and had succeeded to transverse the Rocky Mountains and establish a population east thereof. In the early years of the twenty-first century, the two populations of house finches met in the US Midwest.

The bacterium that caused the fatal eye infection, *Mycoplasma gallisepticum* conjunctivitis, as had been foreseen, then spread to the western finch population, where it was first documented in Missoula, Montana, in 2002 [16] and then in birds caught in Portland, Oregon, in 2004 and 2005. However, the disease didn't have the same rapid spread and devastating results in the western finch population as it had had in the eastern one. A reason for that, as demonstrated in laboratory experiments, was that the genetic variation in the original population in the west was much larger than in the eastern one that emanated from just a few individuals. Genetic diversity was shown to be negatively correlated with infection sensitivity [17].

The house finch is now established all over the USA, Mexico, and South Canada. The total number of birds, in the higher estimates, is assessed to be around 1.4 billion [18].

In a study of how the house sparrow population was influenced, when the house finches first increased explosively in numbers and expanded its area of distribution and then decreased as a result of the mycoplasma infection, the researchers found a strong and clear negative correlation between the two bird species both on aggregated geographic level and on local ones [19]. They concluded that the correlation represented a direct competition, where the finches were the stronger and wasn't just an indirect statistical phenomena.

In this connection, it's worth remembering that when the house sparrows got established in San Francisco in the early 1870s, they didn't for the longest time succeed to spread much outside the city and that California is part of the natural range of the "Hollywood finches."

There are very few published reports of how the number of house sparrows developed in North America during the 25 years from the early 1940s to middle 1960s. In one article that mainly deals with other birds in Canada's Atlantic provinces [20], Christie notes the arrival, explosive number increase, and subsequent decrease and says that the numbers stabilized after the 1930s. In an attempt to explain why the drop in the number of house sparrows stopped, Erskine, in another article [21], points to a partial and temporary return of horses during World War II, when fuel and rubber for tires were reserved for the military. However, winter bird counts in Nova Scotia did not show any clear trend during the first two decades after the war, between 1948 and 1968 [22].

From the end of the 1960s and onward, the decrease in house sparrow numbers is dramatic, and the data on which estimates are done are far better. In Canada's maritime Atlantic provinces, the numbers dropped 90% in cities and 70% in rural areas from the 1960s to early twenty-first century [21].

Also on a far larger geographical scale, there are good numbers from the USA and Canada for the time after the mid-1960s. According to North American Breeding Bird Survey, in which birds annually are systematically counted on about 3000 stretches each some 40 km long, the house sparrow shows a continuing decline for the period 1966–2011 with an average annual −3.8% in the USA and −4.1% in Canada [23]. Accumulated over the 45-year period, this corresponds to a decrease of about five-sixths.

If one takes Wings' figure of 150 million house sparrows in the beginning of the 1940s as a starting point and assumes that the figure remained more or less the same till the mid-1960s and thereafter decreased with about 4% per year, this would give a current population of some 25 million, that is, some 5 percent of what it was at the peak a stiff century ago. And the trend is still down.

There are, however, areas—also major cities—where the dramatic decline in house sparrow numbers has not been evident. I have myself visited Washington DC regularly for more than four decades and often stayed with friends living close to the zoo there. My own observations indicate that the house sparrow still is as abundant in the US capital as it was in the 1970s.

How Do the House Sparrows Fare in Their Native Countries?

This chapter is about the house sparrow as an invasive species in North America. However, having concluded that after an explosive increase in numbers and expansion of territory there during the first half century after introduction, there has been a continuous decline for more than 100 years; it's close at hand to ask how the sparrows have fared lately in England, India, and other parts of the birds' countries of origin. The short answer is "badly."

In Great Britain, the number of house sparrows has dropped with two-thirds since 1970. In an analysis of the development up to 2003, Summer-Smith divides the house sparrow population into three habitat groups with different decline curves. In agricultural areas, the numbers dropped with 60% between the late 1970s and mid-1990s and have stabilized since. In larger cities, there was a steady, slow decline from the 1930s to the turn of the decade 1980s/1990s, when city populations entered a phase of free fall leading to a virtual absence of house sparrows from, e.g., Central London, Edinburgh, and Glasgow. In towns and suburbia, there has for a long time been a steady decline, but not (yet) a catastrophic free fall in numbers [24]. At the time of Summer-Smith's report, 2003, the number of house sparrows in Great Britain was counted to be 6 million. In State of the UK's Birds report from 2012, the number was given as 5.2 million. Increased scarcity of insects to feed the young ones after hatching and increased predation from cats and predatory birds are seen as important factors behind the ongoing decline. Since 2002, the house sparrow is on the UK red list of endangered species.

From India, there are reports of a similar picture with the house sparrows all but absent from major cities [25]. In the rural areas, the decline in the last decades varies from 20% in Kerala, Gujarat, and Rajasthan to 80% in Andhra Pradesh and most coastal regions.

In Sweden, the number of house sparrows is now just below a million as compared to two and a half million 40 years ago, a drop of 60%.

Some Loose Treads (Starlings and Predatory Birds)

In Moseley's tale about the development of different bird species in Ohio [14], he sees a connection between the drop in the number of house sparrows and the raise of starlings, sparrow hawks, and screech owls and interprets a cause-effect relationship. Competition from the starlings, primarily for nesting sites but also in part for food, and predation from owls and hawks pressed the house sparrows back according to him. It's a reasonable hypothesis, but it needs to be supported by more than a negative correlation between the species in an area at one point in time.

On an overreaching level, the number of predatory birds dropped all over North America, as well as in Europe, during the late nineteenth and early twentieth century as a consequence of hunting (and better rifles), and during this time, the explosive increase in the number of house sparrows took place. Later, in the 1950s and 1960s, DDT and other chlorinated insecticides caused another deep decline in the number of predatory birds, of which the sparrow hawk was one. This didn't lead to any increase in the number of sparrows but could have contributed to the stabilization of the house sparrow population after the decrease during the decades before. (The sparrows themselves could of course also have been victims of the chlorinated hydrocarbons but no such effects on populations of the species have been reported.) After the ban on DDT in the 1970s, most predatory bird species, now generally also protected, have recovered, but not all. Several predatory bird species have also established themselves in cities. Thus, on the large scale, there is a reasonable negative correlation between house sparrows and predatory birds, a correlation that also holds true in Europe.

However, it can also be noted that after the period of Moseley's observations, especially after 1966, when the North American Breeding Bird Survey have given a much better picture, the eastern screech owl as well as the sparrow hawk (if Moseley with the term meant *Falco sparverius*) have both declined somewhat in Ohio. It can be argued that this might have happened as a result of less available food, when the house sparrow population shrank, but in the last half century in Ohio, the predators and prey have all declined.

Concerning the interactions between house sparrows and starlings, Andersson [26] concludes referring, e.g., to a study by Weitzel in Nevada that starlings at high population densities can outcompete sparrows for nesting sites that suit both species. In summary though, he states "that there's very little to support the notion that intra-species competition with starlings causes the decline of the house sparrows."

Eurasian tree sparrow (Photo from Wikimedia)

The Eurasian Tree Sparrow (*Passer montanus*) as Comparison

The tree sparrow is a close relative to the house sparrow and at first sight something of a "look-alike." It has a wide distribution in Europe and Asia, which to a large extent overlaps with that of the house sparrow. Also feeding habits overlap, with seeds as a dominant food component most of the year and insects as special hatchling treat.

The Eurasian tree sparrow was introduced into Lafayette Park, Saint Louis, Missouri, in April 1870. A bird dealer got 24 of them in a delivery supposed to contain German songbirds and released them as their song was unlikely to attract any customers. As the birds came from Germany, the species is often referred to as "German sparrow," while the house sparrow is sometimes called "English sparrow" [27].

The species got established in St. Louis and slowly spread to a larger area around the city and continued along the Mississippi river. Today, the distribution reaches from the eastern part of Middle Missouri to Western Illinois and southern Iowa. It's assumed that the tree sparrows "hitch-hiked" med grain carrying river barges as it would explain the current distribution, and the species otherwise shows no special preference for near-water habitats.

The area of distribution today is limited to 48,000 km^2, and the US population consists of some 100,000 individuals. It has thus not had the explosive population development and colonization success of the house sparrow, and the difference between the two closely related species with regard to invasiveness has for long been of interest to researchers. However, no generally accepted explanation has been found besides the more aggressive behavior of the house sparrow. Sometimes, the "homeliness" of the tree sparrow, the tendency of young sparrows to build their own nest close the site where they were hatched, is mentioned as one explanation for the slow spread. Other explanations have to do with by-chance phenomena like that the house sparrows that were brought over had a larger genetic variation than the tree sparrows and thereby a better starting point for natural selection to bring forth individuals adopted to the new environment. The many imports of house sparrows as compared to the one known for the tree sparrows would then contribute to such a difference in genetic variability. Other explanations have to do with diseases and parasites, the idea being that the house sparrows more than the tree sparrows should have escaped theirs, e.g., not brought them to North America to the same extent as the tree sparrows happened to do.

On the other hand, the house sparrow has been a successful invasive species in many parts of the world, while the tree sparrow which was also brought there never reached pest status. That the same by-chance factors should have operated everywhere doesn't seem very likely [28].

An Interesting Sidetrack

The question about the role of diseases and parasites, when introduced species become invasive, is interesting. The simplest explanation, involving pathogens, when a species is relatively more successful in a new environment than in the old, is the one discussed above: the species arrived without (some of) its tormentors and thus fare far better. As far as house sparrow goes, it was observed fairly early on that one of its common parasite in Europe and Asia, the small sparrow flee (*Ceratophyllus fringillae*), didn't make it over the Atlantic [29].

A study of ectoparasites on house sparrows in Europe and North America gives a fuller picture of the situation [30]. In the groups Acarina, Mallophaga, and Siphonoptera, there are a total of 74 species that are house sparrow parasites. Of these, 60 are present in Europe and 37 in North America, while 23 species are present in both sparrow populations and thus, in all likelihood, travelled with the introduced sparrows. The 14 species of these groups that occur as house sparrow parasites only in North America must have come from other birds on that continent. Thus, the sparrows "escaped" 37 of their 60 old parasite species but were hit by 14 new ones.

A different theoretical explanation why the immigrants succeed to establish themselves so well is that they bring diseases and parasites with them, pathogens which they themselves more or less tolerate and can live with but which are lethal to the unaccustomed natives. An example of that from humans is the invasion of Europeans in the Americas.

A study of house sparrows in Kenya, a country the species invaded fairly recently, was set up to illuminate this question but resulted in an unexpected answer: the reason for the invasive success is the ability of the immune system to adjust to new challenges [31].

Summary

After the introduction in the 1850s, the house sparrow quickly spread over North America, and 40 years later, it had colonized large parts of the continent. Then, at its peak, the number could have touched half a billion. After that, the house sparrow population has shrunk. Periodically, this has been a rapid decrease. In between, the numbers have stabilized, before the decline continued. Lately, subpopulation especially in the big cities has all but collapsed. A series of causes have contributed to this long-term, strong decline. The replacement of horses with cars and tractors, the return of many species of predatory birds, and the competition from the native invasive house finch are seen as some of the most important factors. Today, the number of house sparrows in North America is only some 25 million and the trend is downward.

The species doesn't constitute a significant problem anymore.

References

1. Robbins, C. S. (1973). Introduction, spread and present abundance of the house sparrow in North America. *Ornithological Monographs, 14*(14), 3–9.
2. Moulton, M. P., Cropper Jr., W. P., Avery, M. L., & Moulton, L. E. (2010). The earliest house sparrow introductions to North America. *Biological Invasions, 12*(9), 2955–2958.
3. Hooper, J. (1854). Entry for March 7, 1854 Assembly 150.Transactions of the American Institute of New York City 1853. pp. 466–470.
4. Coates, P. (2006). *American perceptions of immigrant and invasive species: Strangers on the land* (p. 256). Berkeley, CA: University of California Press. (Reviewed by Brendon Larsson in Biol Invasions (2008) 10:257–258.)
5. Barrows, W. B. (1889). *The English sparrow (Passer domesticus) in North America, especially in its relation to agriculture*. U.S. Department of Agriculture, Division of Economic Ornithology and Mammalogy Bulletin 1, pp. 1–606.
6. Wing, L. (1943). Spread of the Starling and English sparrow. *Auk, 60*, 74–87.
7. Sauer, J. R., Hines, J. E., Fallon, J. E., Pardieck, K. L., Ziolkowski Jr., D. J., & Link, W. A. (2012). *The North American breeding bird survey, results and analysis 1966–2011*. Laurel, MD: USGS Patuxent Wildlife Research Center.
8. Dunn, R. (2012). The story of the most common bird in the world. Why we love what is rare and despise what is all around us? *Smithsonian Magazine*. Retrieved March 2, 2012, from http://www.smithsonianmag.com/science-nature/the-story-of-the-most-common-bird-in-the-world-113046500/#ixzz2s4t4HItp
9. Fitzwater, W. D. (1994). *House sparrows biology*. New Mexico: Internet Center for Wildlife Damage Management.
10. Pimentel, D. (2002). Chapter 17.4.2: Vertebrate pest. In *Biological invasions: Economic and environmental costs of alien plant, animal and microbe species*. Boca Raton, FL: CRC Press.
11. Pimentel, D., Lach, L., Zuniga, R., & Morrison, D. (2000). Environmental and economic costs associated with non-indigenous species in the United States. *BioScience, 50*(1), 53–65.
12. House Sparrow Damage Assessment. (2008, February 5). *Extension. America's research-based learning network*. Retrieved from http://www.extension.org/pages/11391/house-sparrow-damage-assessment

13. Weber, W. J. (1979). *Health hazards from pigeons, starlings and English sparrows: Diseases and parasites associated with pigeons, starlings, and English sparrows which affect domestic animals*. Fresno, CA: Thomson Publications.
14. Wilson, J. D. (2001). *Common birds of North America—An expanded guidebook*. Minocqua, WI: Willow Creek Press.
15. Moseley, E. L. (1947). Variations in the bird population of the north-central states due to climatic and other changes. *The Auk, 64*(1), 15–35.
16. Badyaev, A. V., & Hill, G. E. (2002, June). Avian quick-change artists, exemplars of rapid adaptation, house finches show that mothers know best. *Natural History Magazine*.
17. Ley, D. H., Scheaffer, D. S., & Dhondt, A. A. (2006). Furthe western spread of mycoplasma gallisepticum infection of house finches. *Journal of Wildlife Diseases, 42*(2), 429–431.
18. Hawley, D. M., Sydenstricker, K. V., Kollias, G. V., & Dhondt, A. A. (2005). Genetic diversity predicts pathogen resistance and cell-mediated immunocompetence in house finches. *Biology Letters, 1*(3), 326–329.
19. House Finch, Life History, All About Birds—Cornell Lab of Ornithology.
20. Cooper, C. B., Hochachka, W. M., & Dhondt, A. A. (2007). Contrasting natural experiment confirm competition between house finches and house sparrows. *Ecology, 88*(4), 864–870.
21. Christie, D. S. (1979). Changes in maritime bird populations, 1878–1978. *Journal of the New Brunswick Museum, 1979*, 132–146.
22. Erskine, A. J. (2006). Recent declines of house sparrows, *Passer domesticus*, in Canada's maritime provinces. *Canadian Field-Naturalist, 120*(1), 43–49.
23. Erskine, J. S. (1968). *Winter birds of Wolfville, Nova Scotia: 1948–1968*. Nova Scotia Museum Occasional Paper Number 7, Science Series (5). 15p.
24. Summers-Smith, J. D. (2003). Decline of the house sparrow: A review. *British Birds, 96*, 439–446.
25. Dandapat, A., Banerjee, D., & Chakraborty, D. (2010). The case of the disappearing house sparrow (*Passer domesticus indicus*). *Veterinary World, 3*(2), 97–100.
26. Ted, R. (2010). *Anderson, biology of the ubiquitous house sparrow: From genes to populations*. New York: Oxford University Press.
27. Dick, G. O. (2004–2014). European tree sparrow. In *Guide to birds of North America*.
28. Ehrlich, P. R., Dobkin, D. S., & Wheye, D. (1988). *Avian inviders*. Retrieved from www.stanford.edu/group/stanfordbirds/text/essays/Avian-Invaders.html
29. Rothschild, M., & Clay, T. (1957). *Fleas, flukes and cuckoos*. MacMillan: A study of bird parasites. New York.
30. Hengeveld, R. (1989, December). *Dynamics of biological invasions* (176p.). New York: Chapman & Hall. (Based on data from Brown NS and Wilson GI (1975), Am. Mdld Natur 94, 154–165).
31. Martin, L. B., Coon, C. A. C., Liebl, A. L., & Schrey, A. W. (2014). Surveillance for microbes and range expansion in house sparrows. *Proceedings of the Royal Society B, 281*(1774), 20132690.

Rabbits in Australia

A classical example and probably the most well-known case of an invasive species is the European rabbit (*Oryctolagus cuniculus*) in Australia. Most people with even a brief interest in biology can retell tales about the devastation to crops and wildlife, that the rabbits brought about; the control attempts with the introduced foxes, that eat easy-to catch kangaroos instead of rabbits; and a gigantic continent-stretching fence, that took so long time to build that the rabbits were equally numerous on both sides, when it was finished. Many may also add that finally a virus disease was introduced and that more or less brought the problem under control.

These popular tales are not untrue, but there is more to the story than that.

The European Rabbit

The rabbit is a well-known small mammalian related to hares. The length of the head and body is 40–55 cm, and they weigh from 1.2 to 2.5 kg as adults. Their coat is mostly grayish with some brown but ranges from sandy yellow to black. The belly and the underside of the tail are white. Among domesticated rabbits also totally white ones are common.

Rabbits are mostly dusk-dawn active and spend much of the day in their often extensive burrow systems called "warrens." Such systems are shared by a group of some 20–40 adults of both sexes, each maintaining a strict hierarchy. The rabbits are born blind and naked, up to seven in a litter, after a gestation period of about a month. They grow fast and sexual maturity is reached after less than 4 months.

For their burrows, rabbits prefer well-drained areas with loose soil with cover of scrub and soil stabilizing rocks. Ideal habitats include small pastures with hedges or stands of bushes and sand dunes. Less liked are conifer forests, damp areas, and areas above the tree line.

A. Jernelöv, *The Long-Term Fate of Invasive Species*,
DOI 10.1007/978-3-319-55396-2_6

The rabbits are herbivores, but as such "omni," accepting a wide range of grasses, herbs, and tree bark. Their natural enemies are any predator large enough to catch them, including foxes, minks, cats, dogs, eagles, and hawks. The young ones are also often eaten by weasel badgers and buzzards.

One of the first areas to which humans brought rabbits was the British Isles. Some sources say the Romans brought them in AD 43 [1], referring to Marcus Terentius Varro (116–27 BC) who wrote that the legions brought rabbits from Spain, while others have it that the Normans took them there in twelfth century [2].

In Britain, they are now considered "native."

The European rabbit (*Oryctolagus cuniculus*)

Introduction and Release in Australia

On board the 11 ships with convicts, marines, and some family members, which landed in Australia in 1788 and was later to be called "the first fleet," were also rabbits. These were domesticated "tame" rabbits, and it's unclear to what extent, if any, these were forefathers to those causing the invasion. That fame is mostly attributed to the 24 wild rabbits that a man named Thomas Austin had sent to Australia and released on his 11,700 ha large property called Barwon Park on Christmas Day 1859. This was in Victoria just outside Geelong. This was at the time of the gold rush and when Australian wool production exploded. Fortunes were made and the lucky rich ones copied the habits of their English peers, and hunting for, e.g., rabbits was one of those sports. Thomas Austin did not only introduce rabbits to Australia. He proudly wrote that he had also introduced hares, blackbirds, thrushes, and partridges [3]. When he could see that the rabbits fared well on his estate, he counted the introduction as one of his accomplishments and invited English gentry to hunt in the "so game-rich area as one seldom sees in England." The duke of Edinburgh was one of those who came in the 1860s. He reportedly succeeded in shooting 416 in only 4 h.

The Duke of Edinburgh rabbit shooting at Barwon Park, Victoria, in the 1860s (Photo from Wikipedia)

Many other farmers followed the example of Thomas Austin and released their rabbits although those were domesticated breeds. A decade after the introduction, the rabbit population was already so large, that a yearly toll through trapping and shooting of two million rabbits didn't noticeably halt the spread.

However, when Thomas Austin released his rabbits in 1859, a wild population was obviously already established in Tasmania. Already in 1827 a newspaper there [4] in an article noted "...the common rabbit is becoming so numerous throughout the colony, that they are running about on some large estates by thousands."

Landowners at the time largely saw the rabbits as a positive addition to the local fauna bringing both meat and fur as well as the pleasure of hunting. If the "success" with rabbit introduction in Tasmania in the 1820s and 1830s prompted anyone to bring individuals of the feral Tasmanian stock over the Bass Strait to Victoria and the Australian mainland is not known, but given the spirit among colonizers at the time, it doesn't seem unlikely that someone tried. Sightings of rabbits at the Eyre Peninsula were reported in the 1830s but not officially confirmed [5].

Regardless of the source of inspiration and stock, the idea of rabbit farming in large enclosures had occurred to settlers also around Sidney in New South Wales. Peter Miller Cunningham, an MD and surgeon-superintendent on convict transports from England to Australia, in a series of letters, later published, noted that "rabbits are bred around houses, but we have yet no wild ones in enclosures." He also wrote that "the scrubby, sandy soil between Sidney and Botany Bay would be ideal for rabbits" [6].

The first warren for extensive rabbit farming in the Sidney area was built by Alexander Macleay at his mansion at Elizabeth Bay House [7]. He was the Colonial Secretary of New South Wales and had arrived in 1826. He was also a distinguished scientist, specializing in entomology, a fellow of Royal Society and Linnean Society, and had served as secretary of the latter. Since 1813 he was also a foreign member of the Royal Swedish Academy of Sciences.

Also Macleay's rabbits fared well on Australian soil, and others were inspired to follow suit and construct stone wall surrounded by warrens of some considerable size. By the 1840s, this was a common practice, and rabbits had become a staple in the diet of ordinary people. Court records from the time show, though, that rabbits were still worth stealing.

Further, there are reports also of rabbit releases near Kapunda. Some sources say in 1866, other in 1870. It could of course have happened there more than once.

The Population Explosion and Spread

Regardless if the origin was only Thomas Austin's wild rabbits from Geelong or if feral domesticated rabbits from other escapes and releases contributed to the population, the rabbit immigrants multiplied and spread quickly. As the start was in the southeast, the rabbits extended their area of occurrence to the north and west. By 1881 wild rabbits overran New South Wales, in 1886 they were seen in South Australia, by 1887 they reached the southwest border of Queensland, and in 1894 they were first seen in the Northern Territories. Overall, the northern spread of the feral rabbits occurred with an annual speed of about 100 km. The westward expansion wasn't quite as fast, but impressive nevertheless. It took till the early 1900s for the rabbits to concur Western Australia. Thus, in less than 50 years, the rabbits colonized all of the Australian continents except the tropical north, some areas of coastal forest, and hot and dry desert.

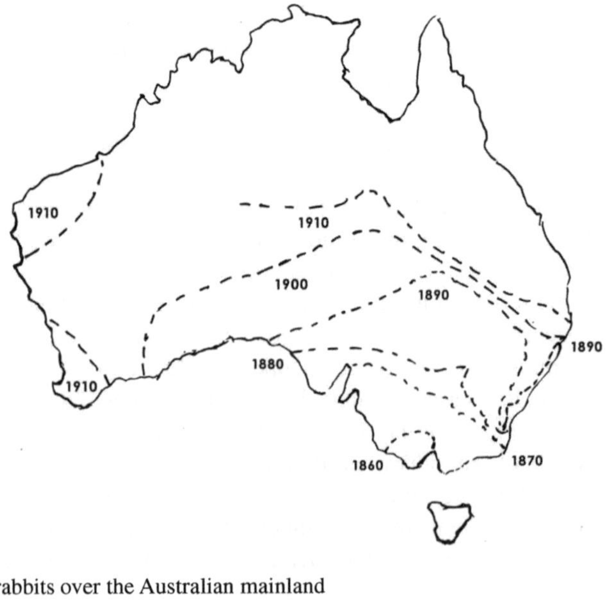

The spread of rabbits over the Australian mainland

During this expansion of its territory, the number of rabbits increased tremendously, but the population estimates are uncertain at most early points in time. CSIRO, the Commonwealth Scientific and Industrial Research Organisation, has the peak number at ten billion in 1926, and this is the most quoted figure [8, 9]. If this is correct, the collective weight of the rabbits at the time would be some 20 billion kg, compared to the 5.4 million humans with a collective weight of perhaps 200 million kg that is 100 times as much. If so, the rabbit population also dwindled the sheep population, which at the time, not having yet recovered from the reduction during the 1902 drought, counted less than 100 million heads and would have weighted around 2 billion kg. In fact there would have been more rabbits, also by weight, than all human and domestic animals put together.

Rabbits around a water hole (Photo from Wikipedia)

Be it as it may with this figure of ten billion in 1920, in the late 1940s, before myxomatosis came into the picture and allegedly became the first effective rabbit control method, the population was estimated to be in the range of 600 million (CSIRO). Should both figures be correct, the rabbit population must have suffered a close to collapse; a 94% reduction in the three decades between 1920 and 1950, despite the fact that none of the control attempts tried during the period, was seen to have any noticeable effect. Surely, if this happened through nature's own control mechanisms, it would be worth knowing more about it. (The easiest explanation is, of course, that the ten billion figures was a gross overestimate.)

The subsequent development of the rabbit population in Australia is a story of fairly successful biological control, of which more will be told below. The intentional introduction of myxomatosis resulted in very high mortality among the infected individuals and reduced the rabbit population to well below 100 million, with some parts of the country, those where the mosquito that spread the virus thrived, almost rabbit free. Gradually, the rabbits became somewhat resistant, even if the disease still kills 40–50% of those infected. The number of rabbits increased again, passing 200 million around 1990, till 1995, when the introduction of the rabbit calicivirus that causes a hemorrhagic disease (RHDV) again brought down the rabbit numbers to low levels, with greatest impact in arid zones and lesser impact in high-rainfall areas.

Surprisingly enough, there seems to be no official information of the current (end of 2013) number of wild and feral rabbits in Australia. Estimates span an order of magnitude from 20 to 200 million. Maybe 50 million is a reasonable guess.

An indication of a historic (since well over 100 years) low number of wild rabbits available for hunters is that commercial rabbit farming, that was banned in Australia for a long time, has become a fast-growing small-scale agro-enterprise during the last 15 years.

Other signs of historic low number of rabbits are smaller populations of foxes, which pre-RHDV to some 60% fed on rabbits [10], and proliferation of woody scrubs like cypress pine, hop bush, and sticky daisy bush in some areas [11].

Ecological Effects

The rabbits are vegetarians and consume a broad variety of plants from grass and herbs to leaf and bark of wooden shrubs. In high densities the rabbits will consume native plants to such an extent that topsoil is bared and the soil exposed to sheet, gully, and wind erosion.

Rabbits reduced Phillip Island, NSW, to a wasteland (Photo from SEWPaC)

But recovery was spectacular after the rabbits were eradicated (Photo from SEWPaC)

Although rabbits are seen as a major factor behind the changing face of the Australian landscape, their exact role in the loss of plant species is more difficult to assess. The reason is that the rabbits didn't come alone. Extensive sheep farming mostly preceded them and many other alien species including feral camels/dromedaries proceeded them. If one, for the sake of discussion, disregard feed preferences among herbivores and compere them with regrd to biomass only, the rabbits would have been larger plant consumers than sheep at the hight of their numbers, during the first half of last century, but not before and thereafter. (Ten rabbits will consume roughly the same amount as one sheep.) However, when it comes to the question of restoring vegetation, rabbits are a bigger problem than domestic animals as they are much more difficult to keep out of any chosen zone. Only on a few islands has it been possible to fully eradicate the rabbits.

The rabbits have also been a strong factor in the demise of a number of small native plant-feeding animals and the reduction in numbers of many more, but again their competition for food resources has not been the only factor in play. The house mouse (*Mus musculus domesticus*) is also an invasive species in Australia occurring over practically the whole continent and is also an important food competitor to small native species. In the agricultural areas in southeastern of Australia, the house mice numbers can reach plague proportions. These mass occurrences have been even more frequent and severe lately, with 2011/2012 setting a record in agricultural damage. It's been suggested that the relative control of the rabbit population in the last decades has given more room—food—for the mice. Another plant-eating feral species is the goat. In 1993, the population was estimated to 2.6 million animals. The numbers fluctuate heavily between years, but with no significant long-term trend since then. Feral goats do a lot of environmental damages but is also seen as having a good economic potential for meat export [12].

Other invasive species such as foxes and feral cats predate on the small native animals, but also on rabbits and mice. It's been a matter of discussions and studies if the drastic reduction in the numbers of rabbits following the introduction of RHDV have benefitted or harmed small native grazers. Obviously, on the one hand, the food competition is reduced with far fewer rabbits around. Obviously also, the predators can rely less on rabbits for food and will increase the predation pressure on the native species. After some time the number of predators would drop, but is it better for the native species to be a larger part of the prey of a small predator population than a small part of the food for a larger group of predators? To make matters even more complicated, there are then also the native predators like dingo, Tasmanian devil, and wedge-tailed eagles that compete both with invasive predators and each other and that eat both invasive and native prey. So far neither mathematical modeling nor field studies have provided any general answer, and often the results of the two do not match. The outcomes of the field studies are highly variable and site specific.

A strong factor and perhaps the real reason for this is man's continuous involvement and interference. Not only is the rabbit population controlled as described above, but farmers use lots of toxins in periods of mouse plague and some of that effect also non-target native grazers. Also foxes and feral cats are subject to control attempts, included with poison, by authorities and landowners. Dingoes and eagles are still shot, although nowadays illegally, as they as seem as threats to newborn lambs.

Thus, the whole situation is one, where not only the rabbit population is controlled but where the whole system of predator-prey and invasive versus native species relationship is manipulated. No real surprise then, that ecosystem models make poor predictions of outcomes, and that field studies yield time and site specific rather than more general results.

Economic Effects

There are a number of estimates of the economic damage to Australia that the rabbit invasion has cost. They use different methods and different reference years. Some of them only include direct damage of the type insurance companies would consider. Others add "the unused potential"; the profit farmers could have had if they had not been forced to reduce operation because of the rabbits. Yet other estimates bring greenhouse gases into the picture calculating carbon dioxide release or lack of fixation into the equation at carbon credit prices. No surprise then that outcomes differ. An example is Bomford and Hart [13] in 2001 estimated the rabbit damage to cost at least A$ 200 million annually. Some years earlier a consultant company, Acil, working for the sheep farmers came up with a figure of A$ 600 million [14].

The Pest Animal Control Cooperative Research Centre (CRC) estimated the direct cost of rabbits to the Australian economy at $113.11 million. Another study by the same organization put the annual loss in economic surplus to Australia due to rabbits at some A$ 200 million [15].

However, CSIRO seems to rely more on the estimates of the Acil company and summarized the situation in the following way in October 2011: "Economic damage by wild rabbits in Australia, including cost of control and production losses, has been estimated at around A$600 million annually. The accumulated damage to our environment over the last 100 years is incalculable but runs into billions of dollars."

In a recent article in Australian Economic History Review, Cooke et al. estimated that biological control of rabbits produced a benefit of A$70 billion (2011 A$ terms) for agricultural industries over the last 60 years [16].

The economic advantages are even more difficult to assess although it's clear that they are small compared to the damage. Already in the 1870s, several million rabbits were shot and trapped every year, and much of the pelts were sold and the meat eaten. During the depression in the 1890s, farmers, station hands, and stockmen often relied on rabbit trapping for food and some extra income and the same became the case during the depression of the 1930s. Also during the two world wars, rabbits provided an important food source fall back for a large part of the poor rural population. Export of rabbit fur, felt hats, and frozen rabbit meat was also periodically of some importance although most of the rabbit products were consumed domestically [17]. For example, in 1944, 104 million rabbit skins/carcasses were exported from Australia [18].

A load of rabbit skins, Northern Tablelands, New South Wales (Photo from Wikipedia)

Control Attempts

Already in the 1870s, when rabbits had spread beyond the limits of Barwon Park, it was noted that livestock production on some pastoral lands fell as much as 60%. The first control attempts simply involved increased shooting and trapping.

When that didn't help, poisoning with, e.g., strychnine-laced grain, became a common practice, followed by fumigation of the burrows and dynamiting of warrens. Today, the most commonly used toxin is "1080" (sodium fluoroacetate) and pindone. Bulldozers have overtaken dynamite as the preferred way to mechanically destroy burrows and warrens.

These methods could lead to some sort of control, but they were laborious and costly and few landowners bordered to keep them up in the long run. To employ them only now and then, when local rabbit populations became particularly large, didn't solve the problem. Some more advanced solution had to be found.

That was also what an Inter-Colonial Commission set out to do in 1890, when it offered £25,000 to anyone who could come up with a solution. This was an enormous sum of money at the time, and a large number of solutions were proposed, most of them judged to be unpractical or based on superstition. One of those that was tried and failed, but anyhow pointed forward, was proposed by Louis Pasteur, the famed French microbiologist and chemist. He suggested that a bacterial disease, chicken cholera, which he tested and found killed also rabbits, could be the answer. A nephew and coworker of his, Adrienne Loire, was sent to Australia with vials of the bacteria (*Pasteurella multocida*). Although highly toxic to rabbits, it was found useless for control of the population, as it had to be injected into each animal and didn't spread from one rabbit to the other.

Foxes

It was well known, probably since prehistoric times, that predators reduce prey populations and many of the proposals the Commission got hinged on that. In Britain, foxes (*Vulpes vulpes*) were one of the well-known predators on rabbits, so it was more or less self-evident that proposals would come, suggesting that foxes should be introduced in Australia to curb the explosion of the rabbit population. By that time, in 1890, however, foxes were already present in Australia. They were not among Thomas Austin prided himself with having introduced, but they came to the same part of the continent, Geelong, Victoria, at about the same time, 1860, and for the same purpose, sport hunting. There are even reports of an introduction in the Sidney area of New South Wales as early as 1845.

When the rabbits spread, the foxes followed suit within a few years, but it seems that without rabbits, the foxes advanced only slowly.

Undoubtedly, the foxes preyed on many other things than rabbits, as they did also in Britain, and some of the smaller native animals were severely reduced in numbers or even driven to extinction. The foxes, however, did also prey on rabbits. Different studies put the rabbit proportion of fox stomach content at between 30 and 60%. Later, experimental fox control programs often showed substantial increases in rabbit populations, e.g., studies of Bank et al. [19]. Subsequent reintroduction of foxes resulted in a return to earlier relatively low numbers of rabbits, but not everywhere [20].

In a larger review of existing studies of the fox-rabbit relationship in Australia, Robley et al. point at differences in response dependent both on rabbit density and geographic area/climatic conditions. They also highlight in their study the role of feral cats as predators of rabbits and competitors to foxes for prey, both with regard to rabbits and to native wildlife [21].

Thus, in relation to the "common knowledge" as mentioned in the vignette and the proposals to the Inter-Colonial Commission of 1890, foxes were not really introduced to Australia to control rabbits, but once both were there, they preyed on them. (This, of course, does not rule out early private initiative to transplant the fox for rabbit control to new areas of Australia, like the later, 1990s, fox introduction to Tasmania [22].

Foxes are now seen as (one of) the most damaging invasive species, and the Australian Federal Government spends more than the double amount yearly to control them as it does with rabbits. The total number of them is estimated at 7.2 million [23].

Rabbit-Proof Fences

In the 1890s, rabbits had reached pest proportions in New South Wales, despite a poisoning program that yearly killed hundreds of thousands.

After a field survey in 1896 rabbit reports from southeast Western Australia, a proposal was made of a Rabbit-Proof Fence [24] for several 100 miles along the state border. Five years later, when a Royal Commission had reviewed the situation and the rabbits had spread further, the decision was to construct a barrier fence across the state from Starvation Harbor on the South Coast to a point near Cape Keraudren on the Northwest Coast. When finished in 1907, it stretched 1834 km and was considered the longest line of unbroken fence in the world. It was called the West Australian No. 1 Rabbit-Proof Fence.

The fence consisted of a wire netting reaching 1 m above ground and 6 in. below. It was supported by posts 12 ft apart, standing 4 ft above ground, and sunk a foot and 9 in. below. Plain and barb wires further stabilized the construction. Later more wires were added on top to make the fence also fox and dingo proof.

Once constructed, the fence was checked and maintained by boundary rider. Initially they used bikes, but later camels (dromedaries) were ridden and used as pack animals.

Long before the fence was completed, however, rabbits were numerous west of it. This resulted in a decision to build a second fence to stem the advance of the rabbits [25]. This fence was given high priority, and although construction of it started 3 years after the first rabbit fence, the No. 2 Rabbit-Proof Fence was finished in 1905, 2 years before No. 1. It stretched for 1166 km from Point Ann on the South Coast through Cunderdin, 150 km east of Perth; the new fence joined the original fence line at Gum Creek in the Murchison area.

In his annual report for the year July 1, 1904, to June 30, 1905, the acting under secretary for agriculture reported that the second fence was completed from the coast to Gum Creek. He also noted that the rabbits had advanced also north of the first fence and recommended the immediate construction of a third fence running west from a place he called Beedeinna to the coast, to protect the northern agricultural areas.

Started in 1906 and finished in 1907, the No. 3 Rabbit-Proof Fence was constructed. It was somewhat simpler than the two others in that the fence posts were further apart and that only two plain wires were used as horizontal support. It did have a barbwire on top, though, and the fence was slightly higher. It ran for 280 km from a point east of Warra Warra and terminated at Bluff Point, about 70 miles north of Geraldton, on the Indian Ocean.

The maintenance of the fences turned out to be a challenging task. Factors of nature as well as man contributed to the problem. The occasional floods in otherwise dry and flat areas were a major destructive force. After dry spells, deadwood from trees and bushes had accumulated, and when heavy rain came, floods swept them away and piled the debris onto the fence till it flattened out or was carried away. In the flat landscape, this could happen over stretches of several kilometers.

Sand drift, especially in the north where part of the fence was built over sand dunes, would pile up sand that completely covered the fence. In some places, a second, third, and even fourth fence had to be built on top of the covered ones.

Bushfires often destroyed the poles, when these were made of wood. Firebreaks, 10 m wide on both sides of the fence, were gradually cleared and the problem diminished.

In every 15 km, or more frequent in more populated areas, gates in the fence allowed passage of people and goods. Very often those gates were left open.

Especially in the early days of No. 2 fence, hordes of rabbits counting 100 or more would come up toward the fence and be forced to turn back. For the landowners on the rabbit side, these forced-to-stop rabbits-on-the-move were like locust swarms. To get rid of them, the landowners often lifted up the fence to allow the rabbit passage under it.

The "gentry" hunters also caused a lot of fence damage. They would cause bigger game like large kangaroos or emus to smash netting and wires, tearing and breaking them. Bullets fired along the net could also cause damage as did certainly out-of-control hunting vehicles that smashed into fence and broke poles and wires.

With time, motorized vehicles replaced camels as a means of transport for the boundary riders, but those were always shorthanded for the task for economic or recruitment reasons. During war times even iron to replace netting, wires and poles would be in very short supply. All in all, the fences deteriorated [25].

Today, only small parts of them still exist and then not for rabbit control, but because they found other uses.

The effect that the Rabbit-Proof Fences had on the spread of rabbits must be concluded to have been small. Arguably, the second fence delayed the westward move of the rabbits with a decade or two.

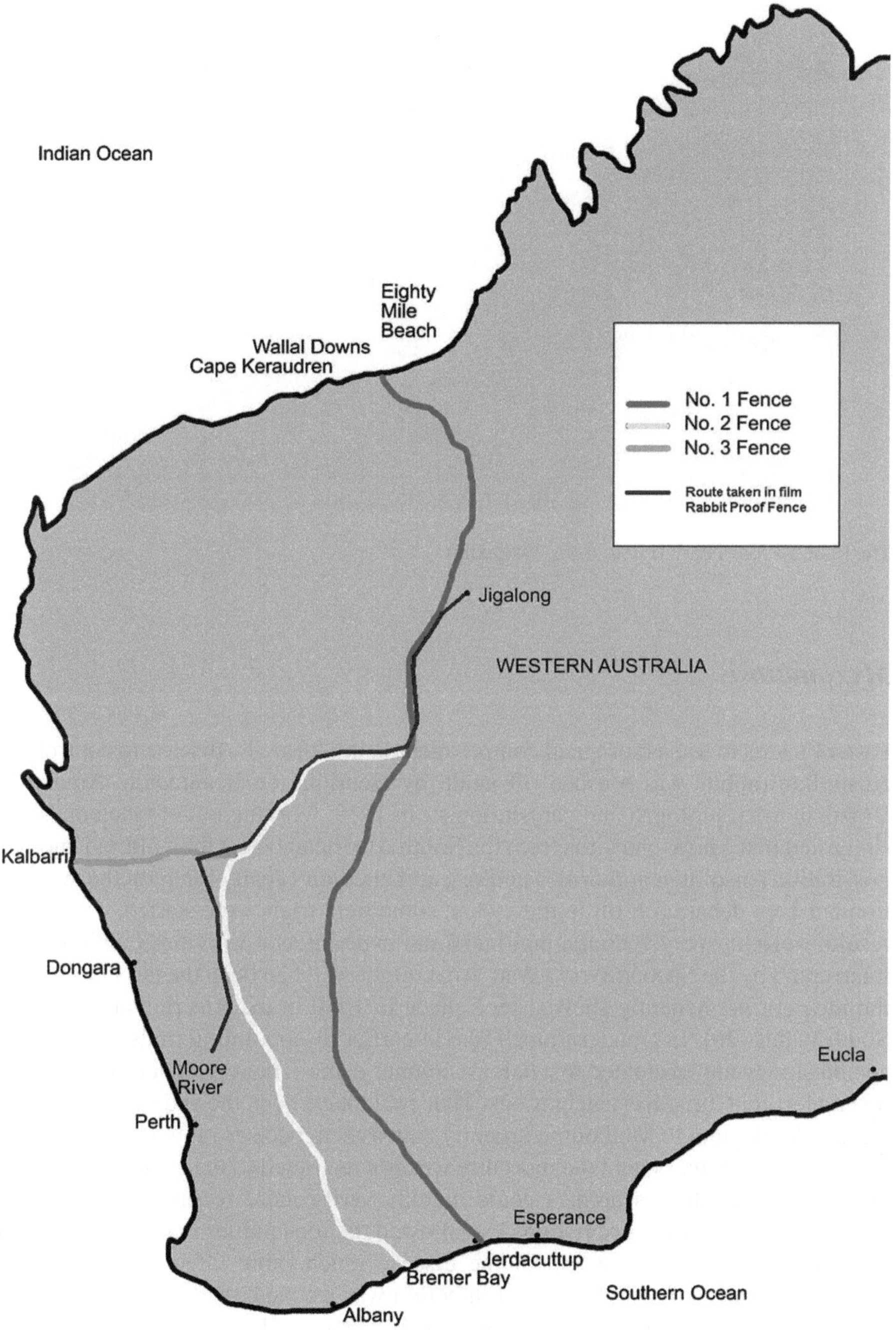

Position of the Rabbit-Proof Fences

The Rabbit-Proof Fence (Photo from Wikipedia)

Myxomatosis

Pasteur's idea to use a biological control agent in the form of a disease to control the Australian rabbits was brushed off again by Henrique de Beaurepaire Aragâo, a Brazilian microbiologist and parasitologist. In 1919, working out of Guatemala, he suggested that a myxoma virus from the South American bush rabbit offered another possibility. Australian authorities and researchers didn't really jump on the suggestion but kept debating it till in the 1930s; some field trials were started. The initial results were not very encouraging, but some experimental activities continued till interrupted by the Second World War. What might well had been the last tests before the idea got permanently shelved took place in 1950 in the Riverina area of New South Wales [26]. In a pattern much like in earlier disappointing trials, the disease spread slowly and grounded to a halt as summer came. The researchers all but gave up and left for Christmas celebrations. That could have been the end of the story but wasn't. A telegram to Melbourne urgently called them back as rabbits suddenly had started to dye en mass, and the mortality was not restricted to the experimental site. On their return, the researchers could quickly verify those reports and found the explanation. Unusually heavy rains had allowed the mosquitoes to multiply beyond their normal numbers and spread the disease much more effectively than ever before, and once a threshold level of infection was reached, there was no stopping. The results were astonishing. The number of rabbits fell with perhaps 90%, and Australia's rabbit problem appeared resolved at the snap of the fingers.

However, over time rabbits developed some degree of resistance, and today somewhat more than half of those infected survive, compared to less than one percent 60 years ago. At times and places, when and where mosquitoes were rare, the spread of myxomatosis was also ineffective. So rabbit population bounced back here and there and other control methods had to be used off and on, but the longer trend was again nevertheless up.

In order to improve the spread of myxomatosis in areas with less mosquitoes, CSIRO made several introductions of the European rabbit flea (*Spilopsyllus cuniculi*), e.g., in 1966. This gave the disease another boast in some temperate areas but was not as effective as one had hoped in the arid zones, to which the fleas were not well adopted. The Spanish rabbit flea (*Xenopsylla cunicularis Smit*), on the other hand, was found to be more so, and in 1993 CSIRO imported also this parasite.

This flea extended the spread of myxomatosis inland, but the effect was dwarfed by that of the RHD virus introduced in 1995.

All in all, the myxoma virus was a game changer in the control of the Australian rabbit, when it was introduced, and it remains a key component in the rabbit control tool kit today.

Rabbit Hemorrhagic Disease

RHD is caused by a calicivirus. It was first detected in Chinese rabbit farms in 1984 and first quickly spread among domesticated as well as wild rabbits over the Eurasian continent [27]. It's an acute disease that results in liver, intestinal, and lymphoid necrosis and a terminal massive clotting of blood in vessels. The symptoms, seldom observed in wild populations where normally just dead rabbits are observed, are lethargy, depression, loss of appetite, lack of coordination, and signs of pain. Nasal froth and discharge may be blood strained.

In Australia, it was introduced in 1995 through a laboratory accident on Wardang Island, when infected rabbits broke out. Initially, the researchers tried to contain the virus by a rabbit extermination program in the vicinity of the breakout point, but, when that failed, they decided to see it as an introduced control agent. Within 8 weeks, the new virus had killed an estimated ten million rabbits.

As it turned out, RHDV pairs well with the myxoma virus in rabbit control, as it's most effective in arid areas, where the latter spreads less well. Although there are signs of a resistance buildup, the initial mortality has dropped from very high to high; it's still rather effective. Furthermore, more virulent strains from Eurasia are being unidentified and gradually imported.

Current Control Status

The big picture today is that the rabbit population is under control, mostly with the help of the two viruses, but with some conventional control methods temporarily and locally needed. The situation, however, appears somewhat precarious and requires continuous surveillance and occasional interference.

What if...?

A Side Story About the Dingoes

The dingo (*Canis lupus dingo*) has ha fairly wide distribution over Southeast Asia. They most likely came to Australia some 4000 years ago as a semidomesticated dog. Analysis of mitochondrial DNA suggests that the Australian population derived from very few individuals, possibly just a pregnant female. They became feral and spread over the continent. In the process they caused or contributed to the extinction on the Australian mainland of previously dominating predators like the Tasmanian devil (*Sarcophilus harrisii*) and Tasmanian tiger (*Thylacinus cynocephalus)*. Fossil records suggest that these two species disappeared some 3000 years ago from the mainland, but remained on Tasmania, where the dingoes were absent.

The dingoes are Australia's largest predator. They live in small flocks or individually (especially younger males) hunt a large variety of preys. Obviously, they were an invasive species, when they arrived, and must have had an effect on the population of prey species also, not only on their marsupial competitors. Today, they are seen as "native," and efforts are made to protect them, not so much from numerical extinction—the population is estimated to be in the range of close to half a million—but from genetic contamination by modern feral and domestic dogs.

For the dingoes, the rabbits, since their introduction, have become a dominant food source in many parts of the continent, and dingo dens are nowadays typically made in rabbit burrows.

References

1. *BBC NEWS—UK—England—Norfolk—Remains of Roman rabbit uncovered.* Retrieved April 13, 2005, from bbc.co.uk
2. Retrieved January 19, 2017, from http://www.britishwildlifecentre.co.uk/planyourvisit/animals/rabbit.html
3. The acclimatisation society. *The Argus*. 21 April 1864. p. 7.
4. *Colonial Times and Tasmanian Advertiser*. 22 May 1827.
5. Rabbit technical note 1 from The Government of South Australia.
6. Cunningham, P. (1827). *Two years in New South Wales*, Vol. 1, p. 304.
7. Sydney Gazette 28 May 1831.
8. Animals CRC (Cooperative Research Centres). Retrieved from http://www.invasiveanimals.com/invasive-animals/rabits/
9. ABC Science. (2009, April 9). *Australia's battle with the bunny*. Retrieved November 8, 2015, from http://www.abc.net.au/science/articles/2009/04/08/2538860.htm
10. Holden, C., & Mutze, G. (2002). Impact of rabbit hemorrhagic disease on introduced predators in the flinders ranges, South Australia. *Wildlife Research, 29*, 615–626.
11. Sandell, P. R. (2002). Implications of rabbit hemorrhagic disease for the short-time recovery of semi-arid woodland communities in North-Western Victoria. *Wildlife Research, 29*, 591–598.
12. (2006). Retrieved November 8, 2015, from https://wilksa.wordpress.com/
13. Bomford, M., & Hart, Q. (2001). Non-indigenous vertebrates in Australia. In D. Pimental (Ed.), *Environmental and economic costs of alien plant, animal and microbe invasions*. Boca Raton, FL: CRC Press.

14. Acil Economics and Policy Pty Ltd. (1996). *The economic importance of rabbits on agriculture production in Australia*. Report Prepared for the International Wool Secretariat.
15. Gong, W., Sinden, J., Braysher, M., & Jones, R. (2009). *The economic impacts of vertebrate pests in Australia*. Canberra: Invasive Animals Cooperative Research Centre.
16. Cooke, B., Chudleigh, P., Simpson, S., & Saunders, G. (2013). The economic benefits of the biological control of rabbits in Australia, 1950–2011. *Australian Economic History Review, 53*(1), 91–107.
17. *Australian encyclopaedia, Vol. VII*. Grolier Society, Sydney.
18. Retrieved November 8, 2015, from http://www.northernplainscmn.com.au/documents/22-LC0298-rabbits-and-their-impact.pdf
19. Banks, P. B., Dickman, C. R., & Newsome, A. E. (1998). Ecological costs of feral predator control: Foxes and rabbits. *Journal of Wildlife Management, 62*, 766–772.
20. Banks, P. B. (2000). Can foxes regulate rabbit populations? *Journal of Wildlife Management, 64*, 401–406.
21. Robley, A., Reddiex, B., Arthur, T., Pech, R., & Forsyth, D. (2004). *Interactions between feral cats, foxes, native carnivores, and rabbits in Australia*. Melbourne: Arthur Rylah Institute for Environmental Research, Department of Sustainability and Environment.
22. Saunders, G., Lane, C., Harris, S., & Dickman, C. (2006). *Foxes in Tasmania: A Report on an Incursion of an Invasive Species*.
23. *Impacts of feral animals*. Game Council of New South Wales. Retrieved March 29, 2012.
24. Broomhall, F. H. *The longest fence in the world*.
25. Crawford, J. S. *History of the State Vermin Barrier Fences (formally known as Rabbit Proof Fences)*.
26. Myers, K. (1954). Studies in the epidemiology of infectious myxomatosis of rabbits. II. Field experiments, August-November 1950, and the first epizootic of myxomatosis in the riverine plain of South-Eastern Australia. *Journal of Hygiene, 52*(1), 47–59.
27. Moss, S. R., Turner, S. L., Trout, R. C., White, P. J., Hudson, P. J., Desai, A., Armesto, M., Forrester, N. L., & Gould, E. A. (2002). Molecular epidemiology of rabbit haemorrhagic disease virus. *Journal of General Virology, 83*, 2461–2467.

Crayfish Plague in Europe

In most European lakes and rivers, crayfish were ubiquitous up to about 100 years ago, with biomasses often matching or even exceeding those of the dominating fish species. At that time, they were generally not highly prized or even much appreciated as food and were seen as a "poor man's dinner" in places such as Scandinavia. In some areas, however, they were seen as delicacies early on, and the resulting price differences spurred a lively and lucrative trade. Concurrent with early industrialization that led to intense water pollution, numerous episodes with widespread fish and crayfish mortalities were reported.

In 1859 and 1860, extensive crayfish mortality was noted in a lake in Northern Italy [1], and in the coming few years, it spread to a number of nearby lakes. As fish were not affected, and several of the lakes lacked industrial discharges, it was concluded that the die-off was not the result of water pollution, but caused instead by an infectious disease.

The first suspect as a causative agent in the crayfish plague was a bacteria of the *Bacillus* family that was accordingly named *B. pestis astaci* [2]. The idea of a connection between crayfish deaths and presence of the invasive *Elodea canadensis* aquatic plant was entertained by leading scientists for a long time [3]; as late as in 1934, Paulus Schiemenz discussed the matter but concluded that the "wasserpest" was (probably) not the cause of the "krebspest" [4].

In 1903, a fungus of the *Aphanomyces* species was identified in diseased crayfish and was claimed to be the cause of the illness. Inoculated specimens grown in culture became infected and developed the disease [5], but it took until 1934 before *A. astaci* was finally confirmed as the agent causing crayfish plague [6].

A. Jernelöv, *The Long-Term Fate of Invasive Species*,
DOI 10.1007/978-3-319-55396-2_7

Noble crayfish (*Astacus astacus*) (Photo from https://de.wiktionary.org/wiki/Flusskrebs)

The Pathogen

Aphanomyces astaci was regarded as a true fungus at the time when it was identified as the cause of the crayfish plague, but the group to which it belongs, *Oomycete*, fungal-like water molds, is now considered to be protists and is classified together with diatoms and brown algae in a group called *Stramenopiles* [7].

A. astaci lives out its vegetative life in the cuticle of crayfish and produces zoospores that infect other crayfish. Zoospores can survive outside of a crayfish for only a week or two, swimming around searching for a new host. When they find one, they attach to the shell, preferably at the joints of antennae and pereiopods and in the soft tissue of the abdomen, encyst, and germinate to start a new cycle as growing hyphae penetrate the crayfish's cuticle.

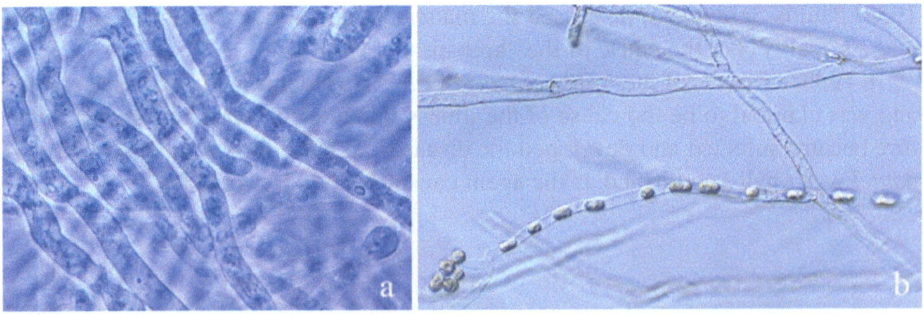

Hyphae (**a**) and sporangium with emerging spore ball (**b**) of the oomycete *Aphanomyces astaci* (Photo Trude Vrålstad (**a**) and David Strand (**b**), Norwegian Veterinary Institute)

In North American crayfish species, *A. astaci* is a fairly benign parasite under normal conditions, as those crayfish are able to delay the penetration by melanin encapsulation and normally shed their shells before the mycelium reaches the soft tissue. Only when conditions are bad and crayfish growth is so slow that shedding is much delayed, the hyphae may have a chance to grow through the shell and to kill the crayfish. European crayfish species, such as *Astacus astacus*, lack this delay mechanism and succumb, when the hyphae penetrate the shell and destroy the tissues beneath.

There is evidence that older, established *A. astaci* strains (type As) that have been present in Europe for many years are less aggressive to the noble crayfish and other indigenous European crayfish species than newer strains (type Ps), introduced with crayfish imports from North America starting in the 1960s.

Signal crayfish, *Pacifastacus leniusculus*, with melanin encapsulation spots (see arrows) showing where the crayfish plague fungus, *Aphanomyces astaci,* has been blocked (Photo from [8])

The Spread

The First Phase

The crayfish mortality seen in Northern Italian lakes in 1859 and onward is generally seen as the first outbreak of the plague in Europe, but there was an earlier mass mortality of crayfish in Lombardi and Treviso that could also well qualify as the premiere [9].

How *Aphanomyces astaci* got there from North America has not been firmly established. Circumstantial indicators have been found to point both at ballast water and imported crustaceans.

The next place with mass mortality of crayfish that came to be seen as caused by the plague was in Eastern-Central France, in lakes and rivers on the Plateau de Langres, starting in 1874 [10]. Most authors have it that the plague spread quickly from there and reached Germany in 1877. Other sources say that the first authentic record of crayfish plague in Germany dates to 1864 [4]. To Austria, it arrived in 1879, Switzerland in 1881, Slovenia in 1885, and Latvia in 1886. With regard to Russia, there are again several different figures, ranging from 1889 [11] to 1892 [12]. To Poland, the crayfish plague came in 1893 and to Finland in that same year [13]. The mode of transfer is thought to have been importation of diseased live crayfish for the food market, contaminated equipment, and boats.

In 1907, the plague reached Sweden with a cargo of crayfish from Finland. At the time, Sweden was a large market for Finnish crayfish export. About 1.9 million of them were shipped in 1906, as the crustacean was moving from being a "poor man's dinner" to the status of a special delicacy for social events. Actually, the fear of importing the crayfish plague had caused Swedish authorities to ban crayfish imports from Finland in 1907—a few weeks too late, it turned out. The captain of the ship that brought the plague found that "the shipment had turned bad" and dumped the dead and dying crayfish in Lake Mälaren outside of Stockholm. Within a year, most of the *Astacus astacus* in the lake were dead [14], and the pest spread to nearby Lake Hjälmaren, the Swedish lake with the largest catch of crayfish at the time—some four million annually—where the fungus also quickly wiped out its hosts.

For two decades, the crayfish plague remained limited to the Mälaren/Hjälmaren lake system, until it turned up in Västergötland in 1928 and then, a few years later, in the crayfish-rich lakes of Småland. There is from there a clear correlation between the spread of the plague and the spread of car ownership, with people transporting contaminated crayfish-catching equipment and boats between water systems. Some of the spread probably also took place through natural means—e.g., by birds dropping parts of dead or dying *Astacus* in other water courses than those in which they had caught them. By 1960, despite efforts to halt the spread of the pest, the indigenous noble crayfish was gone from roughly half of the 30,000 lakes and rivers it had inhabited half a century earlier. Those lakes were the most productive ones, so in terms of catch, they represented a loss of some 90%.

At the time, scientists and administrators had a very pessimistic view of the chances for a comeback of *Astacus astacus*, the native crayfish. It was seen as more or less unavoidable that the plague sooner or later would reach all Swedish crayfish-inhabited bodies of water. Attempts to reintroduce the noble crayfish into lakes Mälaren and Hjälmaren had failed repeatedly, leading to the conclusion that an infected body of water would remain infected or quickly get re-infected should the host species return. Furthermore, it was concluded that it was highly unlikely that the native crayfish would ever develop tolerance of the plague, as no indications of that had been found in 100 years up to that time since the disease had come to Europe and the 50 years since it came to Sweden.

Parallel to the shrinking numbers of waters in which crayfish could be caught, "kräftskivor"—the often raucous crayfish parties beloved of Scandinavians—had advanced to the status of a national August feast, creating a voracious appetite that triggered a huge yearly importation of crayfish. Thereby, the loss of most of the Swedish crayfish fisheries was not only the loss of a favored recreational activity but also a question of the national economy and balance of trade. The Swedish response inspired much of what subsequently happened in most of Europe, and therefore the Swedish story is, by and large, Europe's story.

The Fishery Administration of the Swedish Government (Fiskeristyrelsen) assigned its laboratory for freshwater research with the task of finding a suitable replacement—a species of crayfish tolerant of the plague and as close as possible in taste, external appearance, temperature preference, behavior, and ecology to the noble crayfish. It was clear early on that it had to be a North American species, because that is where the tolerant species lived. Under the leadership of Magnus Fürst, extensive comparative studies of the candidate species were carried out, and in the end it was concluded that the signal crayfish (*Pacifastacus leniusculus*) was the most suitable and its introduction was recommended. Lake Tahoe, in the Californian Sierra Nevada mountains, was chosen as the locality from which to take the specimens to be imported [15]. Tests were made in the small and isolated Lake Träsksjön, and three generations after the introduction of the first signal crayfish the trial was summarized, reporting that the introduced crayfish reproduced and grew well, were ecologically very similar to the noble crayfish and tasted excellent. It was noted that they were strongly resistant to the plague, but could carry the fungi, and that that might be a problem [16]. Based on the positive report, the Swedish Board of Fisheries imported a new batch of Lake Tahoe signal crayfish and introduced them into 60 lakes and rivers in Sweden.

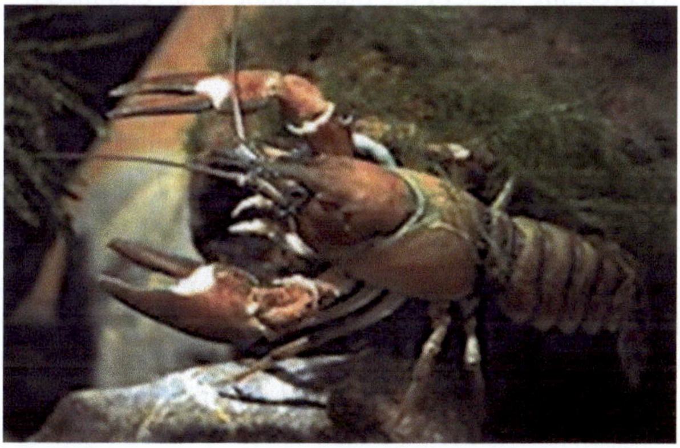

Signal crayfish, *Pacifastacus leniusculus*. The white patches at the joint of the claw fingers is characteristic (Photo from Wikipedia)

The introduction was highly successful, in that the signal crayfish mostly thrived in the bodies of water into which they were introduced. The populations also expanded, so that in less than a decade after introduction to a lake, crayfish catches were back at the old noble crayfish levels or higher. Often, in a classical "overshoot," the populations of signal crayfish, for a while, even became much larger than those of the noble crayfish before the plague. The demand for signal crayfish for introduction was so high that a commercial breeder with high production capacity took over from the state laboratory as brood supplier. At this point, the rest of Europe followed suit, if they hadn't already started their own importation of signal crayfish or other American plague-tolerant crayfish species.

The Second Phase

This started a second phase of the spread of the crayfish plague. The signal crayfish is itself mostly tolerant of the plague and seldom succumbs to it, but, crucially, it is not immune and thus acts as a carrier, infecting the remaining native crayfish as it spreads, thereby also eliminating competition. Mostly, the noble crayfish and other European crayfish species disappear within a year after the arrival of an infected signal crayfish to a watercourse. So in this second phase, the main threat to the remaining native crayfish species in Europe was the signal crayfish and the pest it carried [17].

It may be worth noting that the first signal crayfish from Lake Tahoe might have been free from the infection by *Aphanomyces astaci* despite its general prevalence in North American crayfish populations, although some "fact sheets" now, by omitting the prehistory, seem to imply that the alien signal crayfish brought the plague to Europe and not only spread it once they and the plague were already there. One reason to believe that the signal crayfish did not bring the infection with them is that in a small lake in Finland, which in 1971 was populated with young signal crayfish produced from a breeding stock of the Lake Tahoe strain obtained from Sweden, the noble and the signal crayfish coexisted for 30 years [18–20].

Today, in Sweden, the noble crayfish can be found in less than 1000 bodies of water, a mere 3% of its original distribution [21]. In Western Continental Europe, it is even less, while in Eastern Europe, including Ukraine, Belarus, and Russia, a much higher proportion of lakes and rivers have maintained their noble crayfish populations. In many of the watercourses where they remain, *Astacus astacus* is doing very well, as water pollution and acidification are receding. All in all, however, the global decline (over three generations) is estimated by the IUCN to be 40–50%, so the species is classified as vulnerable [22].

Assisted by the signal crayfish and other introduced American crayfish species, the European plague continued to spread and reached Spain in the late 1960s and Norway, Greece, and Turkey in the 1980s [23]. At the time, Turkey was a major exporter of crayfish to Sweden and other Western European countries, with, for example, over 5000 tons of the Turkish crayfish (*Astacus leptodactylus*) exported in 1984. The plague and overfishing put an end to that, and by 1991, the total catch in the country was only 200 tons [24].

In Britain, where the white-clawed crayfish, *Austropotamobius pallipes*, is the only native crayfish species, the plague seems to have arrived only with the signal crayfish, of which some 300 implants were made from the mid-1970s and onward [25]. The first clear case of crayfish plague was found in the Avon River in 1981, and the first diagnosis of *A. astaci* was made in 1984 [26]. There are, however, a number of reports from much earlier dates, of mortalities and drastic fluctuations in numbers, for which no explanation has been found [27, 28].

Since the arrival of the plague, the number of native crayfish has dwindled, especially in the Thames River catchment area and in Hampshire, where some 95% of the original population are assessed to have gone. The white-clawed crayfish is now classified as endangered by the IUCN, and similar declines in their number have been reported from France and Italy [29].

The plague, however, is not seen as the only cause for the demise of the white-clawed crayfish. Other factors include direct competition from nonnative crayfish species, water pollution, and habitat destruction. Five alien crayfish species are firmly established on the British Isles: Apart from the signal crayfish, there are two other plague-tolerant and plague-dispersing American species, the red swamp (*Procambarus clarkii*) and the spiny-cheek crayfish (*Orconectes limosus*), as well as two European crayfish plague-susceptible species, the noble (*Astacus astacus*) and the Turkish crayfish (*Astacus leptodactylus*). In addition, the marbled crayfish (*Procambarus sp.*), a popular aquarium species, seems to be on its way toward establishing footholds in several watercourses [30].

The alien Turkish crayfish has also fallen victim to the plague in its new British home [31], and the noble crayfish is obviously risking the same fate—if so, to the applause of EU and native species purists.

Ecological Effects

Noble crayfish have a major impact on the aquatic systems in which they are found, especially, of course, when living in dense populations. As bottom-living consumers near the top of the food chain, they consume other bottom-living animals such as mussels and snails, other crustaceans, worms, and insect larvae. They also eat plant material and significantly reduce the number of rooted macrophytes and attached algae, in addition to often being the dominant feeders of detritus. This means that they push the lakes in which they live toward an ecological balance, wherein phytoplankton are the dominant primary producers, and zooplankton and plankton-eating fish constitute much of the next trophic levels. Several fish species eat crayfish, especially when they are small or shedding, perch and eel being among the most important predators. Several species of waterfowl, otter, and mink also have crayfish on their menus. In dense populations, the main predator of young crayfish may be older cohorts of the same species.

As the crayfish plague killed off the European crayfish species in the century between the 1860s and the 1960s, the change pushed the limnic system to develop more macrophytic plants and a more diverse bottom fauna. Most of this, however, was hidden by a steep increase in nutrient discharges leading to eutrophication, algal blooms, and anoxic bottoms. From the 1960s and onward, however, when signal crayfish entered the European scene and took over the empty niche left by the noble crayfish and other European crayfish species, the reversal of this change was also to a large extent overlooked. Measures against water pollution then lead to a gradually decreasing nutrient discharge and a nutrient cycle push in the opposite direction.

When the signal crayfish itself became the main vector of the plague, little ecological change took place as the new species quickly replaced the old one, the ecological characteristics of which are very similar.

Economic Effects

The economic effects of the crayfish plague, like the spread of the disease and its ecological effects, cannot be discussed without considering the role of the imported signal crayfish.

The estimated catch of crayfish in Sweden—100% noble crayfish at the time—has been estimated to have been some 1000 tons annually before the plague hit in 1907 [32]. The annual consumption of crayfish in Sweden today is around 6000 tons, of which some 4000 tons are imported. (In 2016, the importation from China exceeded 3000 tons, which triggered a Chinese export tariff of 20% and sent importers looking for other sources [33]. The figures, even the official ones estimating crayfish catches in Sweden, vary a lot. According to Magnus Fürst [15] referring to "Fakta om Svenskt Fiske," the 2004–2005 catch was 2663 tons. The official statistics in Statistiska Meddelanden [34, 35] puts the figure at 189 tons caught by professional fishermen in 2015 and 589 tons by recreational fishermen in the same year. Others claim more generally that some 1000–1500 tons of crayfish are caught annually in Sweden [36], but note that the figure is based on questionnaires and is thus uncertain.

Estimates of the proportions of signal and noble crayfish species also vary, with resulting annual catch estimates of *Astacus astacus* to be in the 50–150 tons range. It thus seems that although the catch of noble crayfish has dropped 85–95%, the catch of signal crayfish has more than compensated for that loss, both in weight and also, more or less, in value.

The situation in the rest of Europe is similar. Economically, then, the costs of the plague can be measured in lost revenue from the drop in crayfish catches between the 1860s and 1980s in continental Europe and from 1907 to, say, 1977 in Sweden. A rough estimate of the former is 120 years × 5000 tons/year = 600,000 tons, and the latter would be 70,000 tons. At today's significantly higher price for crayfish, that would amount to $15 billion and $1.7 billion, respectively.

To these figures should, of course, be added both the costs incurred in trying to prevent the spread of the plague and those of introducing the signal crayfish (by many environmentalists today seen as a catastrophic decision resulting in the creation of a new invasive species), bringing the total historic sum for Europe as a whole, to somewhere in the ballpark of $20 billion at today's prices.

Control Attempts

As it was deemed impossible from the start to try to cure the infected crayfish by killing the oomycetes on the living hosts, attempts to control the disease have been directed to reducing the spread by banning movement of crayfish specimens from infected areas to uninfected ones and by requiring that catch equipment, boats, and other objects that have been in infected waters are carefully dried or washed in formalin or other disinfecting solutions before they are brought to uninfected bodies of water. Naturally, such rules and regulations are difficult to control, and success has been inconsistent.

Since the signal crayfish arrived and took over as the main vector of the plague's spread, control efforts have focused on preventing its further spread. In many parts of Europe, people interested in catching crayfish have not adhered to the above policies, so unauthorized or illegal implants have been a main factor behind the continued spread of the crayfish plague.

There have been many attempts to eradicate different, nonnative crayfish species in several European countries. In a review of the methods used or proposed [37], a range of mechanical, biological, physical, and chemical methods are evaluated, including the use of biocides and pheromones. Mechanical methods such as trapping, seining, and pond drainage have been found to be able to reduce, but not eliminate, crayfish populations. Biological control, for instance, by introducing large numbers of European eel, *Anguilla anguilla*, was seen as promising, but there have been no known examples of successful extermination by increased fish predation [38].

In some cases, in smaller bodies of water, extermination field trials have been successful. In one such successful trial in Scotland, natural pyrethrum was used [39]. In Norway, where the first signal crayfish was identified in a small pond in the Dammane area of Telemark County in 2006 [40], there had been an attempt to eradicate them with a combination of the synthetic pyrethroid cypermethrin (commercial formula BETAMAX VET.®) and pond drainage in 2008, the initial results of which were reported to have been positive [41].

Treatment of a Norwegian pond with BETAMAX VET.® using a pump to disperse the chemical (Photo by Roar Sandodden) [41]

On the Virtues or Lack Thereof of Introducing the Signal Crayfish to Europe

Ecologically and economically, the signal crayfish has largely taken over the role of the noble crayfish, especially in Western Europe and Scandinavia. Despite this, the critique of the decision to introduce the signal crayfish following the destruction caused by the crayfish plague has been increasing in volume within the scientific community engaged in working with invasive species. This is in line with a general trend to focus more on preservation of native species and the original biodiversity and less on production and profitability. Two further, very specific, reasons for the reevaluation of the feasibility of the signal crayfish are also often mentioned: the first concerns the question of whether the noble crayfish, despite the consensus reached by the experts 50 years ago, has in fact belatedly started to develop some resistance to the plague. Discovery of *A. astacus* survivors in lakes in which the plague wiped out the real majority of crayfish is seen as indication of such an adaptation. The fact that the virulence of the first imported strain of *Aphanomyces astaci*, the As strain, today is much lower and more variable than that of the post-1960 imports, the Ps strain, is another indication [42]. (The implication of this interpretation is that what took place was more as an adaptation of the fungus than of the crayfish). The keyword, though, is "today." Was the As strain of the plague more aggressive 100 or 150 years ago than it is today? If so, then adaptation is likely to be underway, and introduction of the signal crayfish and the Ps strain of the plague might have set the process back a century or so. If not, then Europe just happened to get a less aggressive strain of the plague in the first import (but one that was, even so, destructive enough virtually to eliminate European crayfish species from tens of thousands of lakes across the continent), while other American strains, perhaps from other crayfish species that came later, were even worse.

The second reason for reevaluation of the introduction of signal crayfish is that although the average yield of crayfish is higher from *Pacifastacus leniusculus* than from *Astacus astacus*, the variation is also higher. The signal crayfish produces more but smaller eggs, leading to higher fecundity when the conditions for survival are good, but to much larger losses when they are not. Thus, year classes vary very much in size, with years with cold springs being particularly bad in Northern Europe. The smaller larvae of the signal crayfish are also more exposed to predation of fishes, including many in the carp family, and to cannibalism of older conspecifics. With the perception headwind blowing against the signal crayfish, these variations in population size and catches have received much more attention lately, and their long-term higher yield compared to the noble crayfish has been called into question. Several scholars have also speculated that signal crayfish population crashes such as the one in Lake Mälaren, Sweden, in the first years of this millennium, could be caused by a virus that interacts with the plague defense mechanisms of the signal crayfish, rendering them sensitive to the disease [43]. Others see the extensive introduction to the lake during that time of baby eels, or elvers—up to 30 tons or some 20 million individuals in some years—as a more near-at-hand explanation [15].

The accumulated economic value of the signal crayfish catch in Europe since its introduction might be in the range of $5–10 billion, based on a Swedish figure of $1–2 billion. This more or less offsets the value of the lost noble crayfish catch during the decades since its introduction, and had it not been introduced, these sums would have had to be added to the overall total cost of the crayfish plague over the last century.

Among alternatives to the signal crayfish, it is sometimes suggested that the genes behind its tolerance mechanism, the melanin encapsulation that hinders and delays the growth of the *A. astaci* hyphae, could be transferred to the noble crayfish, thereby creating a plague-tolerant GMO version of the species [44].

Official Policies and Public Perception

All over Europe, the official policy today is to limit the spread of the crayfish plague and, wherever possible, to restore populations of native crayfish species. As the signal crayfish, besides being an alien itself, is also the main vector of the disease, the policy focus is on them, and the ambition is to halt their spread and—in some countries—preferably to extinguish them. For a long time now—in Sweden for several decades—no new permits for stocking lakes with signal crayfish have been granted. However, the continued illegal stocking of private waters has continued, both spreading the plague and illustrating the widespread public disagreement with the restrictive governmental policies. These, however, are now being taken another step forward. As of August 3, 2016, according to an EU directive, it is "prohibited to barter, breed, raise, transport and use the [signal crayfish] species," and as of August 3, 2017, it is also prohibited—together with 36 other species on the list of invasive aliens—to sell them [45].

In Sweden, these directives have raised significant public concern both as to the future of their crayfish parties and to the equally social catching of crayfish, reflecting the fact that the public generally views the crayfish very positively and does not particularly discriminate between the noble and the signal crayfish species. This concern about the possible loss of a beloved tradition has been raised repeatedly in the Parliament, and the promulgation of national rules and regulations to ensure compliance with the EU directives are awaited both with anxiety (from those loving to catch and eat crayfish) and expectations (from scientists in the invasion field and environmental groups) [45]. In most other European countries, in which there is no tradition of crayfish parties, opinions are less divided.

References

1. Cornalia, E. (1860). Sulla malattia dei gamberi. *Atti della Società Italiana di Scienze Naturali, II*, 334–336.
2. Hofer, B. (1898). Über die krebspest. *Allgemeine Fischerei-Zeitung, 23*, 293–300.

3. Seligo, A. (1895). Bemerkungen über krebspest, wasserpest, Lebensverhältnisse des Krebses. *Zeitschrift Fischerei, 3*, 247–261.
4. *Aphanomyces astaci*—iLumina Digital Collection. UNCW Randall Library.
5. Schikora, F. (1903). Über die Krebspest und ihren Erreger. *Fischerei Zeitung, 6*, 353–355.
6. Schäperclaus, W. (1935). [English title not available]. (Die Ursache der pestartigen Krebsterben.) *Zeitschrift für Fischerei, 33*, 343–366.
7. Manual of Diagnostic Tests for Aquatic Animals. (2009). Chapter 2.2.1.—Crayfish plague (*Aphanomyces astaci*).
8. Diéguez-Uribeondo, J., Temino, C., & Müzquiz, J. L. (1997). The Crayfish Plague Fungus (*Aphanomyces astaci*) in Spain. *Bulletin Français de la Pêche et de la Pisciculture, 347*, 753–763.
9. Ninni, A. P. (1865). Sulla mortalita dei gambari (Astacus fluviatilis L.) nel veneto e piu particalarmenta nella provincia trevigiana. *Antonelli, 10*, 1203–1209.
10. Raveret-Wattel, M. C. (1885). Résumé des résponses au questionnaire Sur la maladie des écrevisses. *Bulletin of Society Acclimation, 4*, 614–615.
11. Retrieved from http://www.cabi.org/isc/datasheet/93177
12. Arnold, J. (1900). Kurzer Bericht über die Verbreitung der krebspest in Russland und über den gegenwärtigen Zustand des Krebsfanges in dem Wolgagebiet. *Allgemeine Fischerei-Zeitung, 25*, 449.
13. Viljamaa-Dirks, S. (2016). *Epidemiology of crayfish plague*. Doctoral dissertation, Helsinki University.
14. Alm, G. (1924). Notiser rörande kräftpesten i Sverige. *Svensk Fiskeritidskrift, 33*(6), 165–173.
15. Fürst, M. (2006). *Debatten om signalkräftan: ett främmande skadedjur som borde utrotas eller räddningen av en folkkär tradition?* Retrieved from http://docplayer.se/7960214-Debatten-om-signalkraftan-ett-frammande-skadedjur-som-borde-utrotas-eller-raddningen-av-en-folkkar-tradition.html
16. Svärdson, G. (1965). The American crayfish *Pacifastacus leniusculus* (Dana) introduced into Sweden. *Report of the Institute of Freshwater Research, Drottningholm, 46*, 90–94.
17. Bohman, P., Nordwall, F., & Edsman, L. (2006). The effect of the large-scale introduction of signal crayfish on the spread of crayfish plague in Sweden. *Bulletin Français de la Pêche et de la Pisciculture, 380–381*, 1291–1302.
18. Nylund, V., & Westman, K. (2000). The prevalence of crayfish plague (*Aphanomyces astaci*) in two signal crayfish (*Pacifastacus leniusculus*) populations in Finland. *Journal of Crustacean Biology, 20*(4), 777–785.
19. Westman, K., & Savolainen, R. (2001). Long term study of competition between two co-occurring crayfish species, the native *Astacus astacus* L. and the introduced *Pacifastacus leniusculus* Dana, in a Finnish lake. *Bulletin Français de la Pêche et de la Pisciculture, 361*, 613–627.
20. Westman, K., & Pursiainen, M. (1979). Development of the European crayfish *Astacus astacus* (L.) and American signal crayfish *Pacifastacus leniusculus* (Dana) populations in a small Finnish lake. *Fresh Water Crayfish, 4*, 244–250.
21. Retrieved December 21, 2016, from http://www.raddaflodkraftan.se/historik/
22. Retrieved December 31, 2016, from http://www.iucnredlist.org/details/2191/0
23. Alderman, D. J. (1996). Geographical spread of bacterial and fungal diseases of crustaceans. *Revue Scientifique et Technique, 15*(2), 603–632.
24. Aydin, H., Harlioglu, M., & Deniz, T. (2012). Harvest, export and economic status of freshwater crayfish (Astacus Leptodactylus Esch.1823) in Turkey. *African Journal of Agricultural Research, 7*(16), 2463–2468.
25. Holdich, D. M., & Ree, I. D. (1991). Distribution of freshwater crayfish in the British isles, with particular reference to crayfish plague, alien introductions and water quality. *Aquatic Conservation: Marine and Freshwater Ecosystems, 1*, 139–158.
26. Alderman, D. J., Polglase, J. L., Frayling, M., & Hogger, J. (1994). The crayfish plague in Britain. *Journal of Fish Diseases, 7*(5), 401–405.

27. Cornish, C. (1902). *The living animals of the world*. London: Dodd, Mead & Co.
28. Duffield, J. E. (1933). Fluctuations in numbers among freshwater crayfish, Potamobius pallipes. *Journal of Applied Ecology, 2*(2), 184–196.
29. Retrieved January 3, 2017, from http://www.iucnredlist.org/details/2430/0
30. Retrieved January 3, 2017, from http://www.thefishsite.com/articles/403/nonnative-crayfish-in-the-uk/
31. Retrieved January 3, 2017, from http://news.bbc.co.uk/2/hi/uk_news/england/7546789.stm
32. Fiskeriverket. (1993). *Möjligheter att öka flodkräftbestånd i svenska vatten* (p. 60). Rapport till Jordbruksdepartementet, Drottningholm: Sötvattenslaboratoriet.
33. Retrieved from http://www.di.se/artiklar/2016/7/28/kraftig-prisuppgang-pa-kraftor/
34. JO 56 SM 1601.
35. JO 57 SM 1401.
36. Edsman, L. (2004). The Swedish story about import of live crayfish. *Bulletin Français de la Pêche et de la Pisciculture, 282*(372), 372–373.
37. Holdich, D. M., Gydemo, R., & Rogers, D. W. (1999). A review of possible methods for controlling nuisance populations of alien crayfish. In F. Gherardi & D. M. Holdich (Eds.), *Crustacean issues 11, Crayfish in Europe as alien species* (pp. 245–270). Rotterdam: A. A. Balkema.
38. Ribbens, J. C. H., & Graham, J. L. (2004). *Strategy for the containment and possible eradication of American signal crayfish (Pacifastacus leniusculus) in the River Dee catchment and Skyre Burn catchment, Dumfries and Galloway*. Scottish Natural Heritage Commissioned Report No. 014, pp. 1–51.
39. Peay, S., & Hiley, P. D. (2006). *Biocide trial to eradicate signal crayfish in the North Esk catchment*. Scottish Natural Heritage Commissioned Report No. 0122.
40. Johnsen, S. I., Taugbol, T., Andersen, O., Museth, J., & Vralstad, T. (2007). The first record of the non-indigenous signal crayfish Pasifastacus leniusculus in Norway. *Biological Invasions, 9*, 939–941.
41. Sandodden, R., & Johnsen, S. I. (2010). Eradication of introduced signal crayfish Pasifastacus leniusculus using the pharmaceutical BETAMAX VET.®. *Aquatic Invasions, 5*(1), 75–81.
42. Viljamaa-Dirks, S. *Epidemiology of crayfish plague*. Retrieved from https://helda.helsinki.fi/handle/10138/168313?show=full
43. Söderhäll, K. (2006). *Svenska forskare och företag spred kräftdöden över Europa*. Dagens Nyheter, Dagens debatt, 4 April.
44. Söderhäll, K. (1990). Kräftpest är ett allvarligt hot mot signalkräftan. *Suomen Kalankasvattaja/Fiskodlaren, 5*, 10–11.
45. Retrieved from http://www.aftonbladet.se/nyheter/article23238386.ab

The Colorado (Potato) Beetle

How It Went from Obscurity to Becoming a Plague

The Colorado beetle (*Leptinotarsa decemlineata*) is an eye-catching, easy-to-recognize insect whose cover wings sport ten black stripes against a yellowish background.

This type of coloration is generally a warning signal to potential predators that the species is toxic and should be avoided. This is also the case with regard to the Colorado beetle. Like other species of the genus, it produces leptinotarsin, a neurotoxin that releases acetylcholine at nerve synapses, thereby wreaking havoc on nerve impulses. Leptinotarsins are highly toxic to a wide range of organisms, including vertebrates. A very similar substance to that produced by the Colorado beetle was used as an arrow poison.

The fact that such a striking insect was first noted only in 1811 and scientifically described in 1824 is a clear indication that it was uncommon or even rare at the time. The first observation, by Thomas Nuttal in the Rocky Mountains and on the specimens collected for Thomas Say's entomological studies, was on the Texas thistle (*Solanum rostratum*), one of the tumbleweed-forming plants. Subsequently, larvae and adults were found on a variety of wild solanaceous plants including horse nettle, ground cherry, pepper, and tobacco. After that, it was no great surprise to entomologists when the beetle was found on a potato plant (*Solanum tuberosum*), which first happened in Nebraska in 1859 [1].

The original native range of the species, however, is thought to be central and northern Mexico up to the South Central USA.

The inclusion of the domestic potato as one of its host plants dramatically changed the fortunes of the beetle. It now had vast fields of food at its disposal and little competition, as most insects, snails, and other possible competitors cannot handle solanine, the glycoalkaloid found among nightshade species such as potatoes and tomatoes. (Some scientists have proposed that the toxins found in the

A. Jernelöv, *The Long-Term Fate of Invasive Species*,
DOI 10.1007/978-3-319-55396-2_8

Photo from Wikipedia

Colorado beetle larvae actually are derivatives of the toxins of the plant they are eating.) The beetle made full use of this culinary Eldorado when it opened up and began to mass-multiply. The first place where this caught the eye of the public was in Colorado in 1865. Two years later, the name of the state was firmly attached to the name of the beetle.

From then, the spread across the potato-growing areas of the North American continent was rapid. By 1870 it had reached Detroit, and in the following year, the Detroit River was reported to be literarily covered by inches of Colorado beetles for weeks at the end of summer. An article in *The New York Times* described how trains skidded on rails that had been "oiled" by crushed potato beetles. At that time or shortly thereafter, the beetle crossed the Canadian border to the north; expanding their territory eastward, they reached the Atlantic coast in 1874 [2].

Some Biological Facts About the Colorado Beetle Itself

The Colorado potato beetle is a 1–1.5 cm large member of the Chrysomelidae family. One of its other popular names, ten-striped spearman, describes it by one of its most characteristic features—the ten dark brown stripes, five on each cover wing. The larvae, as well as the adults, feed on plants in the *Solanum* genus. In the spring, a Colorado beetle female will lay her yellow eggs in batches of 20–40 on the underside of the sprouting leaves of the host plant. In total she may produce up to 800 of them.

Eggs and larvae of the Colorado beetle (Photos from Wikipedia)

These hatch into black-headed, red-bodied larvae, which after four developmental stages drop to the ground and dig a decimeter into the soil to pupate. Dependent on temperatures, the entire process from egg to imago takes a month or two.

The process of going from pupa to beetle is also weather dependent. If winter is approaching, the larvae in the pupae stage can go into diapause and delay emergence until spring. Normally, though, the Colorado beetle will overwinter as an adult.

As mentioned above, the Colorado beetle produces its own toxin and may also use that of the host plant for further protection. Consequently, it has relatively few natural predators that feed on it. However, one North American ground beetle, *Lebia grandis*, is a specialist predator on the eggs and larvae of the Colorado beetle, being immune to its toxins. The ground beetle larvae are even obligate parasitoids on the Colorado beetle pupae, meaning that they feed on nothing else.

It is believed that this specialization of *Lebia grandis* on *Leptinotarsa decemlineata* is a recent phenomenon that began only some 150 years ago when the latter had made the potato its prime host plant and started to mass-develop. Earlier, another ground beetle, *Leptinotarsa juncta*, is thought to have been the prey for *Lebia grandis* [3].

Another natural enemy of the Colorado beetle is the pathogenic fungus *Beauveria bassiana*, which infects a wide range of insect species and is now commercially available in different formulations as a biological control agent.

The Colorado Beetle in Europe and Asia

In 1877, the Colorado beetle reached Europe, where its presence was documented in Germany and England. In both countries, eradication programs were successful. Given later, futile, attempts, it is unfortunate that the methods used were not specified, but given the limited alternatives available at the time, uprooting and destroying the potatoes in all nearby fields were most likely to have met with the greatest success.

Given the volume of cross-Atlantic trade, it is not surprising that these first emigrants were followed by more. In 1901, an outbreak devastated the potato crops on some fields in Tilbury, Essex, but the Colorado beetles were again eradicated the following year. A number of such episodes have taken place since, and the beetle has never secured enough of a foothold on the British Isles to become endemic—with the exception of the Isle of Jersey. During the German occupation of Jersey during World War II, the Colorado beetle arrived to and settled on the island [4].

Nowadays, there are some 30–60 British Customs interceptions annually of shipments on which Colorado beetles have been found, mostly on fresh vegetables such as parsley from France and produce from other EU countries [5].

During or just after World War I, the beetle became established on fields near US military bases in Bordeaux, France. During the 1920s, they spread from there to Spain, but mostly they expanded northward. At the onset of the World War II, they had reached Belgium and the Netherlands, and from there, during the 1940s, they progressed east into Germany. During the war and the Cold War that followed, the possible use of the Colorado beetle as an agent of biological warfare became a hot political topic, with associated accusations and denials, as detailed in a later section.

By the last part of the 1940s, the potato beetle was well established in Poland, and several large campaigns for its eradication or control were launched. The first entry into the Soviet Union occurred in 1949, but a forceful eradication effort succeeded and strict control measures put in place were able to keep the beetle out until 1958 [6].

On its further eastward spread, the Colorado beetle entered China in the western province of Xinjiang in 1993 and has since continued its encroachment at the relatively slow speed of some 25 km per year [7].

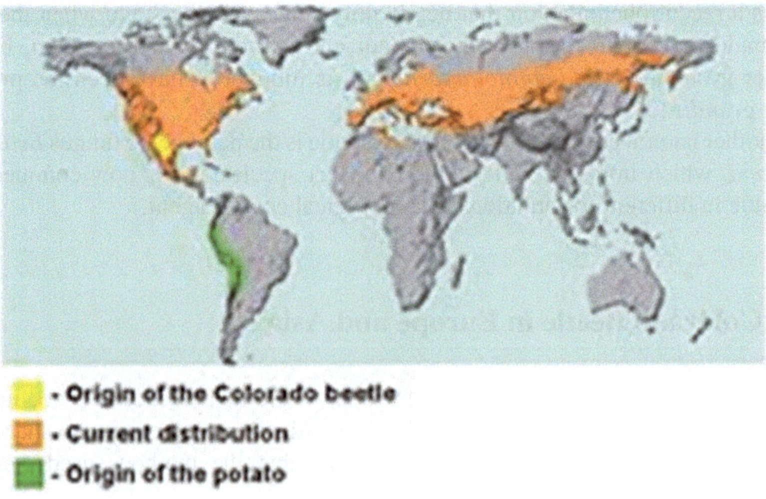

Distribution of the Colorado potato beetle (Wikipedia 2014)

In countries in which the Colorado beetle is not endemic, short-lived infestations may still occur, as it did in Finland in the summer of 2011 when strong winds drove masses of potato beetles into the country from Russia [8].

The Colorado potato beetles in Europe and Asia show a much smaller genetic diversity than their North American relatives, strongly indicating few or even a single successful founding attempt, involving a very limited number of individuals [9].

An Agent of Biological Warfare?

In the summer of 1940, when Germany occupied a large part of France, including Paris, one of the conquered facilities to which they paid special attention housed the laboratories of the Poudrerie Nationale, in Le Bouchet. A German specialist, Professor H. Kliewe, a former director of Giessen University's Diagnostic Laboratory for Infectious Diseases, concluded in a report to the Heeressanitätsinspektion (HSI) (Military Surgeon General) that the laboratory had made advanced preparations for biological warfare involving, among other things, anthrax. Documents from the Le Bouchet laboratory, later captured in Lyon, revealed that the French had indeed seen great military potential in biological agents and that their British colleagues and counterparts shared that assessment [10].

As a response, the Germans set up a division within the HSI, with professor Kliewe as its leader, with the specific task of investigating all aspects of biological warfare. The division also had access to agents operating in England, as well as to their reports. One such report, dated April 30, 1942, talked about a B-24 Liberator aircraft arriving to England from the USA with a cargo of 15,000 Colorado potato beetles and a large number of Texas ticks.

The report leads the Oberkommando des Heeres (OKH) (German Army High Command) to request an assessment from HSI as to what extent Germany would be vulnerable in case of an invasion by either or both of these organisms. In Kliewe's reply, he saw no great danger from the Texas tick but feared that the Colorado beetle could severely damage German potato production and thereby the country's ability to fight a prolonged war. Based on this, the OKH ordered the establishment of a Kartoffelkäferabwehrdienst (Potato Beetle Defence Service) including a Kartoffelkäferforschungsinstitut (Potato Beetle Research Institute) in Kruft to be fully operational before the end of 1942.

The work of the research institute soon enough covered not only the defensive aspects but also the offensive uses of the Colorado beetle. The east coast of England, according to German intelligence estimates, had some 400,000 ha of potato fields that could be targeted. It was calculated that 20–40 million potato beetles would be required to devastate them, and to produce the insects on a large scale, a Colorado potato beetle breeding program was initiated in June 1943 with the aim of having sufficient quantities on hand a year later.

In preparation for these attacks, field trials were carried out to study the effects on dispersal of release height, winds, and temperature. For the purpose, 40,000 live, camouflaged (with paint!) potato beetles were airdropped over fields near Speyer in

October 1943. A second trial, involving 14,000 Colorado beetles, was carried out some months later. On time, according to the schedule he had been given, Kliewe in June 1944 informed the OKH that the experiments had all been concluded and the preparations were complete, so deployment of the Colorado potato beetle would be possible at any time [11].

From military sources and archives, there is nothing to indicate that any German attack on Britain using Colorado beetles ever took place, despite the preparations. However, in a 1979 article in the *International Herald Tribune*, a prominent British naturalist, Richard Ford, claimed to have first-hand knowledge of several Colorado beetle bombings. The first one, according to him, took place in 1943 near Chale, on the Isle of Wight. Teams of children, including himself, were sworn to secrecy and sent to sites of suspected beetle attacks, where they were instructed to pick up the black and yellow bugs and drop them into boiling water to kill them [12].

The laboratories of the Kartoffelkäferforschungsinstitut in Kruft were destroyed by bombs and artillery fire in late 1944. The fate of the purportedly tens of millions of Colorado beetles raised there is not known.

A second round of real or propaganda war involving the Colorado beetle can be seen as having started on May 23, 1950, when a farmer, Max Tröger, in the village Schönfels bei Zwickau in East Germany, spotted an American airplane flying over his village. The following morning, he found his fields covered with potato beetles.

The East German press over the following months reported a number of cases in which local plagues of potato beetles followed overflights of unidentified or American planes. DDR politicians raged against "the six-legged soldiers that led the American invasion," and government reports talked about "the criminal attack by American imperialist warmongers on our people's food supply" [13].

The story did not stop there but continued with similar allegations from the communist governments in Poland and Czechoslovakia that was followed by a large Warsaw Pact campaign called the war against the potato beetle, in which hundreds of thousands of, mostly, teenagers were mobilized to handpick beetles from potato fields [14].

The Americans, of course, denied all allegations and dismissed them as propaganda.

The Damage Inflicted by the Colorado Beetle

Both the adult beetle and its larvae live off the foliage of the potato or other Solanum plant. From the hatching of the egg till the larvae pupates, it will have consumed some 40 cm^2, and the adult will during its average life span eat another 10 cm^2 [15].

An experimental field on which one part has been repeatedly and successfully sprayed with insecticides and the other left untreated (Photo by Gary Sewell)

In larger potato-growing regions such as East Germany in the early 1950s, the beetle might, if not effectively controlled, reduce potato harvests by half; if they were allowed to continue unchecked, they would destroy it totally.

How to Control the Beetle? Mechanical, Chemical, and Biological Attempts

Mechanical, or physical, if one prefers that terminology, control of the Colorado beetle basically takes place in two forms. One is directed toward the insect itself and can be illustrated by the abovementioned examples in which children in England and in Warsaw Pact countries were sent out to potato fields to handpick adults and larvae and submerge them in hot water or other lethal liquids. The other form involves starving the beetles to death by eliminating their food source, that is, by destroying all potato and tomato plants over a sufficiently large area. An alternative to the uprooting and disposal of plants that yields the same end result is to use flamethrowers and burn them in situ.

Uprooting and disposal can be done on a fairly large scale with machines, while so far for individual insect-picking only manual methods are available, if one does not count hand-held vacuum cleaners. Today, mechanical methods are mostly employed only in small-scale gardening, although one can argue that large-scale crop rotation schemes that deliberately exclude *Solanum* plants over several consecutive years achieve the same thing as uprooting and disposing of potato plants but in a less destructive way: they deprive the Colorado potato beetle of its feed.

Chemical control of the Colorado beetle was practiced on a small scale already in the late nineteenth century. The early chemicals of choice were arsenicals, such as Paris green and lead arsenate, and, to some degree, rotenone (Brown 1951). Large-scale chemical control commonly practiced in the late 1940s and early 1950s involved using DDT as the agent. Many scientists and practitioners thought the Colorado beetle problem, together with those of other insect pests, could be solved with DDT. That was also the view of the committee that awarded the Nobel Prize to Paul Hermann Müller in 1948 for the discovery of the insecticidal properties of DDT. However, it only took until 1952 for the first cases of DDT-resistant Colorado beetle to be discovered. The agricultural industry shifted to dieldrin as the anti-beetle insecticide of choice, and by 1958 the Colorado beetle had again mutated to tolerate this chemical as well. A long list of new pesticides followed, but the Colorado beetle showed an astonishing ability quickly to develop resistance to all of them. In some cases, this happened after just a year of use (e.g., with the chemical pesticide endrin), and in at least in one case (that of oxamyl), it occurred during the very first year. Presently, the Colorado beetle is resistant to a wide range of insecticides from different chemical groups—organochlorines, organophosphates, carbamates and pyrethroids, organotins, isoflavones, neonicotinoids, macrocyclic lactones, nereistoxin analogues, and spinosyns, as well as endotoxins of *Bacillus thuringiensis tenebrionis* [16].

A reason for this ability to develop resistance to insecticides is the coevolution of the Colorado beetle and plants in the Solanaceae family, with the toxic glycoalkaloids in their foliage, that required the beetle to evolve the enzymes to detoxify the poisons. With relative small modifications, these detoxification systems have been altered to enable them also to deal with insecticides.

The diagram below, which is based on data from the Arthropod Pesticide Resistance Database, 2007, plots the cumulative number of insecticides to which resistance in the Colorado potato beetle has been reported.

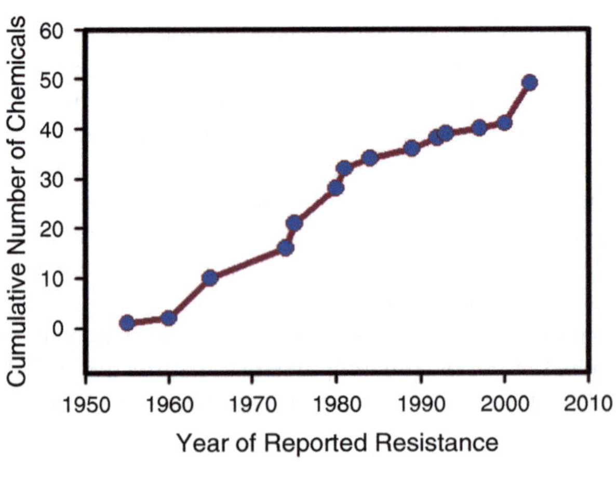

> "North-eastern USA is clearly an area with very severe resistance problems dependent on the agricultural practices there, but insecticide resistance in the Colorado potato beetle is a truly global phenomenon. Obviously, not every beetle population is resistant to every single compound that has ever been observed to fail against this pest. However, both cross-resistance and multiple-resistance appear to be very common within the tested populations. Resistance problems reached critical levels in the U.S. in the early 1990s, when growers in some potato-producing regions completely ran out of chemical control options. Arrival of neo-nicotinoid insecticides in 1995 brought a period of relief in areas where the beetles had developed resistance to other chemicals. However, the first instances of resistance to imidacloprid were soon reported from commercial potato farms in several U.S. states in 2000." [17]

Among bio-pesticides, various products with the active ingredient azadirachtin (from neem tree seeds) have some efficacy against the Colorado beetle in the early larvae stages, but here, as in the case of bacterial toxins from *Bacillus thuringiensis*, resistance has been found in several beetle populations, and this resistance is likely to spread if the applications become more widespread.

One of the first attempts to control the Colorado beetle biologically took place in Illinois in the 1890s, when a group of desperate potato farmers released thousands of chicken onto their fields to roam and—it was hoped—pick off the beetle adults and larvae. There are no reports as to the outcome with regard to the potato harvest or the fate of the chickens, but the fact that the method did not catch on among neighboring fellow farmers and was discontinued where it began suggests that it was not cost-effective.

These days, there is an entire battery of organisms that is used for biological control of the potato beetle, albeit mostly among "green gardeners" rather than commercial potato farmers. To start with, there are natural enemies mentioned above, the ground beetle *Lebia grandis* and the pathogenic fungus *Beauveria bassiana.* There are also two species of flies that act as parasites on the Colorado beetle larvae, *Doryphorophaga doryphore* and *D. coberrans*, and one wasp, *Edovum puttleri*, that goes for the eggs.

Another group of organisms that are promoted as antidotes against the Colorado beetle are entomopathogenic nematodes. Some commercially available species including *Steinernema carpocapsae*, *S. feltiae*, *Heterorhabditis megidis*, *H. marelata*, and *H. bacteriophora* are said to have shown a very high efficacy against adult, larval, and prepupal stages of the Colorado beetle when tested in soil under laboratory conditions [18, 19].

Currently, the methods for Colorado beetle control in commercial potato farming largely involve crop rotation and utilizing still-permitted pesticides from different chemical groups, also on a rotational basis. Under such management regimes, the number of beetles can (mostly) be kept relatively low and the damage to the potato crop within tolerable limits. There seems to be little prospect, however, of eradication with these methods.

In gardening, biological and mechanical control methods are much more in use, but these also do not offer much prospect of extermination of the Colorado potato beetle.

Economic Aspects of the Colorado Beetle Infestation

There are many studies of local and temporal economic losses in connection with major infestations of the Colorado beetle, but there seems to be no comprehensive global assessment. Starting with some guesstimates of overall production and economic value based on national or regional assessments of the treatment costs and net losses in spite of control efforts, a crude idea can be obtained.

Global potato production in the parts of the world that the Colorado beetle has colonized is around 200 million tons annually. The average farm-gate sale price of unprocessed potatoes could be in the range of $100 per ton, leading to an estimated total value of $20 billion. The cost of insect control (here attributed entirely to the Colorado beetle) may be 5% (in the USA 8% of operational expenses) or $1 billion. The production lost to insect damage despite control attempts may be around 10% or another $2 billion. The damage inflicted by the beetle on other commercial crops such as tomatoes and eggplants may add another $500 million or so.

This back-of-an-envelope-calculation leads to an estimated global Colorado beetle damage figure of some $3.5 billion. Excluding North America, where the beetle is considered to be native, the damage, control, and opportunity costs of the invasive potato beetle could be in the range of $2 billion annually.

In China, today the world's largest potato producer, where the potato beetle has not yet entered the major potato-growing areas, current control costs and losses to the beetle in the parts of Xinjiang where it is present has been estimated to be a modest $3.2 million, while the potential damage if/when it has spread to the whole country is given as $235 million [7], also that a surprisingly low figure.

Current Trends and the Most Likely Future

Most likely, the continued spread of the Colorado potato beetle in the next decade or so will include the major potato-growing areas in (northern) China. From there, even if it is a longer leap over lands with little tradition of potato cultivation, there is no real reason to believe that the potato fields in northern India will be spared in the long run. Australia and New Zealand both have climates suitable for the beetle but may stand a somewhat better chance to prevent it from getting a permanent foothold, much as the British Isles have done so far. Areas at risk following climate change are the Nordic countries.

With regard to control in the future, it beggars belief that any of the insecticides in line for market introduction will break the seemingly endless cycle of insecticide-resistance-new insecticide that has been so characteristic of Colorado beetle management to date.

Summary

Since *Leptinotarsa decemlineata* evolved in the 1840s to include the potato among the other *Solanum* species that serve as its hosts, the insect has spread from a limited native area on the Mexican-USA border to occupy a practically circumglobal belt across the North American and Eurasian landmasses. It is not only one of the most successful invading species (and arguably the one causing the most damage to agriculture); it is also one of the organisms that has so far shown the greatest ability to develop resistance to insecticides. There is little doubt that it will continue to spread, eventually covering most of the potato-growing areas of the world, and there seems to be little chance of its eradication.

References

1. University of Florida. (2007). Featured creatures: *Leptinotarsa* spp. Retrieved October 1, 2014.
2. Venturini, G. Retrieved October 1, 2014, from http://www.photomazza.com/?Leptinotarsa-decemlineata
3. Weber, D. C., Rowley, D. L., Greenstone, M. H., & Athanas, M. M. (2006). Prey preference and host suitability of the predatory and parasiroid carabid beetle, *Lebia grandis*, for several species of *Leptinotarsa* beetles. *Journal of Insect Science, 6*(9), 1–14.
4. Retrieved from http://webarchive.nationalarchives.gov.uk/20130123162956/http:/www.defra.gov.uk/evidence/economics/foodfarm/evaluation/planth/annex1_3.pdf
5. Cannon, R. J. C., & Bartlett, P. W. (1998). *Colorado beetle interceptions in the UK*. CSL, MAFF publication, 1-2.
6. Alyokhin, A., Vincent, C., & Giordanengo, P. (Eds.). (2012). *Insect pests of potato: Global perspectives on biology and management* (p. 616). Amsterdam: Academic Press.
7. Liu, N., Li, Y., & Zhang, R. (2012). Invasion of Colorado potato beetle, *Leptinotarsa decemlineata*, in China: Dispersal, occurrence, and economic impact. *Entomologia Experimentalis et Applicata, 143*, 207–217.
8. Koloradonkuoriainen hyppyyttää tarkastaiia perunapelloilla, YLE news (in Finnish).
9. Grapputo, A., Boman, S., Lindström, L., Lyytinen, A., & Mappes, J. (2005). The voyage of an invasive species across continents: Genetic diversity of North American and European Colorado potato beetle populations. *Molecular Ecology, 14*, 4207–4219.
10. Retrieved October 2, 2014, from http://www.sussex.ac.uk/Units/spru/hsp/documents/CWCB33-Garrett.pdf
11. Barnes, J. M., Cromartie, W. J., Henze, C., & Hofer, J. W. (1945, September 12). *A review of German activities in the field of Biological Warfare* (Report No B-C-H-H/305). p. 133. (The report was originally classified 'secret', it was declassified on 17 July 1992 and is available through the US National Archives).
12. International Herald Tribune 25 Feb 1970, p 5, "When the Nazis tried to starve out Britain by beetle-bombing crops". Also sent out by AP from Yarmouth.
13. Burns, L. (2013, September). The great cold war potato beetle battle. *BBC World Service*, 3.
14. Lockwood, J. A. (2008). *Six-legged soldiers: Using insects as weapons of war* (p. 136). New York: Oxford University Press.
15. Ferro, D. N., Logan, J. A., Voss, R. H., & Elkinton, J. S. (1985). Colorado potato beetle (Coleoptera: Chrysomelidae) temperature-dependent growth and feeding rates. *Environmental Entomology, 14*, 343–348.

16. Retrieved October 3, 2014, from http://resistance.potatobeetle.org/History.htm
17. Alyokhin, A., Baker, M., Mota-Sanchez, D., Dively, G., & Grafius, E. (2008). Colorado potato beetle resistance to insecticides. *American Journal of Potato Research, 85*, 395–413.
18. Ebrahimi, L., Niknam, G., & Lewis, E. E. (2011). Lethal and sublethal effects of Iranian isolates of *Steinernema feltiae* and *Heterorhabditis bacteriophora* on the Colorado potato beetle, *Leptinotarsa decemlineata*. *BioControl, 56*, 781–788.
19. Cantelo, W. W., & Nickle, W. R. (1992). Susceptibility of prepupae of the Colorado potato beetle (coleoptera, chrysomelidae) to entomopathogenic nematodes (Rhabditida, Steinernematidae, Heterorhabditidae). *Journal of Entomological Science, 27*, 37–43.

Water Hyacinths in Africa and Asia

"Today, I've seen the most adorable flower in a pond in the garden of Dusit (a palace of Chulalongkorn's). I love it and I just got to have it, but where could I find seeds?" (In a letter to a friend from a wife of a Dutch expert adviser to the Siamese government.)

The Plant

Water hyacinths (*Eichhornia crassipes*) are native to South America. It's been introduced to much of the tropical and subtropical world and is now considered an aquatic pest from the southern states in the USA, via Africa and Asia, to Australia and New Zealand.

The plant itself is a free-floating aquatic macrophyte with broad and thick leaves, ovate in shape, and glossy. They sit on bulbous, spongy stalks and float above the water surface or stick up in the air. The free-hanging roots are dark purple, sometimes almost black in color, and form a dense, fibrous, feathery system. During favorable conditions the plant may be almost a meter high. The flowers sit on a spike supported by an erect stalk. There are 8–15 of them, from purple to pink in color, and quite attractive. That, of course, is a prime reason for their spread as humans have introduced them into garden ponds and similar bodies of water.

The plant is fast growing and can double its biomass in a week. Reproduction is mostly vegetative outside South America, but seeds are occasionally formed and from them plants, maintaining a genetic diversity through recombination of genes. The stands of waterhyazinths often become so dense and thick that they can carry a person—at least on skies.

A. Jernelöv, *The Long-Term Fate of Invasive Species*,
DOI 10.1007/978-3-319-55396-2_9

The water hyacinth, *Eichhornia crassipes* (Photographer Vic Ramey). Photo courtesy UF/IFAS Center for Aquatic and Invasive Plants

Introduction and Spread to Africa and Asia

Where the water hyacinth comes from and where it can be found today are not in question, but how it first got there is less clear and subject to some debate. According to one school of thoughts, the Cotton States Exposition in New Orleans in 1884 played a key role. It's claimed that this was the first time that live plants were brought from the Amazon and shown to the public. Many visitors found the flowers attractive and secured plants from the Japanese pavilion of the expo, and these were then spread to various places in the USA and to London and Paris as well as Amsterdam. Fifteen years later, water hyacinths on the loose had become a pest that clogged waterways throughout the US coastal states [1].

From London, Paris, and Amsterdam, the plant then made it onwards to British, French, and Dutch colonies and areas of influence in Africa and Asia. The Australian continent saw the plant in 1890. In the 1890s it was also found in the Nile delta outside the city of Alexandria [2, 3].

From South Africa water hyacinths on the loose were first reported in some bodies of water on the Cape Flats in 1908 [4, 5] and some lakes in KwaZulu-Natal around 1910 [6].

It's generally said to have reached India in 1888 or 1889 [7], although an earlier date, 1886, of introduction and a direct import from Brazil has also been suggested [8].

To the Dutch East Indies, today's Indonesia, and Bogor, the botanical garden on Java, the water hyacinth found its way in 1894. There the plant grew so fast that within a year cartloads of it were dumped into the Tji Liwung River, which flows through the gardens. From there, probably, it was brought to Japan during the Meiji period at around 1900 and from Java it was also taken to Bangkok in 1901, allegedly by a consort of king Chulalongkorn's, who was taken in by its beauty and stocked the royal garden ponds with it [9, 10]. From there it spread to the Chao Phraya delta and along the Mekong River and reached Laos, Burma (today's Myanmar), Cambodia, and Vietnam. There, however, it was already present. How it first got there is not totally clear, but already in 1902, concerns about mass development and canal clogging were reported [11, 12].

A Different School of Thoughts as to the Spread

According to these ideas, the water hyacinth was brought alive to Europe and hence to Africa earlier and through other routes than the New Orleans exposition of 1884. The famous expedition of Alexander von Humboldt in South America 1799–1804 is in this case seen as a source of water hyacinth seeds and seedlings. Aimé Bonpland, a French botanist and key companion of von Humboldt's, is thought to have sent samples of water hyacinths to empress Josephine, who passed them on to the French botanist Alire Raffeneau-Delile. He was professor of natural history at the University of Montpellier and director of the botanical garden there. By the late 1790s, when France under emperor Napoleon occupied Egypt (1797–1807), the botanical garden of Montpellier had a sort of field station in the Nile valley, as it had on several African islands. Delile is supposed to have grown water hyacinths there already then, a practice continued by the Brits after the departure of the French [13]. For further spread, proponents of this line of thought hold botanical gardens, museums, and catholic priests and monks largely responsible [14–17].

Ecological Effects

When water hyacinths form the dense stands they are capable of under condition of strong light and nutrient-rich water and these floating islands cover areas measured in km^2, the ecological effects are dramatic. Under this cover there is hardly any light, which effectively eliminates both phytoplankton and submerged aquatic plants. As the leaves of the water hyacinth almost all are above or floating on water, the oxygen that the plant produces goes to the atmosphere and not to the water, while dead parts of the plant sink and consume oxygen while degraded. Thus, below the carpet of water hyacinths, the water is often anaerobic or with a low oxygen content. That in turn will have a negative effect on most fish species and other types of aquatic life.

Given the mass of plants and their rapid growth, it's obvious that water hyacinths will all but monopolize available nutrients like nitrogen and phosphorus.

In connection with photosynthesis and growth, water hyacinths will also evapotranspire a lot of water, much more than what would evaporate from an uncovered water surface, and thereby effect water balance, e.g., in dams and water magazines.

That a dense plant cover of large water surface areas also impedes human uses like boating and fishing goes without saying.

These negative effects caused by large, dense stands are not so pronounced when the water hyacinths occur as single plants or smaller floating aggregates. Then positive effects, like the cover and protection that the root system offers to young and small fishes from predators, may outweigh the drawbacks.

Economic Effects

As can be seen from the case studies below, different areas in Africa and Asia have had severe problems with associated high economic costs during different periods of time. Damage and control cost figures are only off and on available on a local or regional basis, but a rough estimate is that for almost a century they seldom exceeded $200 million annually and seldom went below $50 million in today's currency. In total for the century they could have been as high as $10 billion. Today, in Africa and Asia, the yearly costs are in the lower part of the annual range and partly compensated by the value of the usages.

Case Story: Lake Inle

Water hyacinths probably came to Burma in the first decade of the twentieth century, although there are no reports to verify that claim. In the second decade of the century, however, there are several accounts of their presence and the nuisance they cause, and in 1917 a law, the Water Hyacinth Act, was passed. It said: "The presence of the water hyacinth in the Union of Burma is hereby declared to be a public nuisance." Article 4 of the act stipulated: "No person shall possess or keep the water hyacinth (*Eichhornia crassipes*) and every owner or occupier shall destroy any water hyacinth growing in or on any place belonging to or occupied by him." Article 6 stated the punishment for non-conformance: "… shall be guilty of an offence, and shall be liable on conviction thereof to a fine not exceeding one hundred rupees, or upon a second or subsequent conviction to a fine not exceeding five hundred rupees" [18].

Strictly speaking, the people living around Lake Inle, the Inthas, are often in conflict with these legal provisions as most of them at least from time to time possess the plants.

Lake Inle is a fairly large lake in middle Myanmar (Burma). It is known primarily for two things: the way men (not women) row their boats standing up in front of it using one leg for the purpose and for their floating gardens. Zoologists would add the endemic species to the list: There are more than 20 species of snails and at least 9 species of fish that live only here. It's however the gardening that's most relevant with regard to the water hyacinths. When they were first introduced, there was a mass development of them in Lake Inle like in so many other places. Shorelines, shoreside canals, and mouths of rivers going to the lake got clogged, and there were fears that the whole lake would be covered. There was even a local rule for a time that any boat with spare space should fill it up with water hyacinths and bring them to shore for disposal.

However, the Inthas quickly found a better use of the plant. Their traditional floating gardens were created by people who either fenced in a small area of the lake, a few meters wide, with bamboo poles and filled it up with cutout squares of grasses and soil from the marshes around the lake or rooted macrophytes from deeper parts of the lake. Alternatively, they anchored floating plant material that floods tore loose from the inlets and flushed out into the lake by sticking bamboo poles through it. They then kept adding plant material and occasionally bottom sludge till the mixture of decaying floating organic matter was a meter thick, with a quarter or so sticking up above the water surface. Then they planted tomatoes or other vegetables and flowers. The gardens float and will rise and fall with the water level and will never be flooded and never dry out. Water hyacinths became excellent additions to the plant material that formed the gardens. Today it often constitutes the bulk of the construction material.

The sudden increase in the amount of available garden-forming material meant that more of them could be built quicker and easier, and as this coincided with a growing population, the area under culture increased. Today, the floating gardens cover some 20 km^2 mostly in the shallow western part of the lake [19]. Some 100,000 people depend on them and other lake-linked activities like fishing and tourism for their living.

Notes: Floating gardens on Lake Inle (Source: photo taken by Ralf-André Lettau (photo credit, Wikipedia))

This made the water hyacinth popular among the Intha people, who also have resisted suggestions to introduce South American beadles and other biological control agents. The Inthas have also, in accordance with a long tradition of theirs to use fibers of lotus to produce a "silk" and from tree bark to make paper for umbrellas, used dried parts of water hyacinths to produce furniture. These handicrafts and small-scale industries are not unimportant as economic activities.

In summary: Today water hyacinths in Lake Inle are not seen as much of a threat but rather as a resource, at least by the Intha people, the lake dwellers.

Case Story: East Bengal

To Bengal, it's supposed that the plant arrived more than a century ago. A Scottish jute merchant, George Morgan, is credited with the import according to some sources. A different narrative suggests that the Calcutta Botanical Garden got the plant in the 1890s and that English ladies, charmed by the beauty of the flowers, got samples from there and transplanted them into their Dhaka garden ponds. Some also saw it as a German biological warfare against British India, and the name they gave the plant, the "German pana," stuck for a time in the region. Nowadays it's mostly referred to as kachuripana.

By the 1920s it had spread to most bodies of water in the low lands of today's Bangladesh, and when much of the agricultural lands there were flooded during the rainy season, water hyacinths colonized those areas too. Thus, it didn't only obstruct traffic, most of which was water bound, but also interfered with the planting, growing, and harvesting in the wet agriculture including production of, e.g., rice and jute. In these and other ways, such as impeding fishery, the water hyacinths had a significant negative effect on the economy.

The first reaction to those problems was of course to try to remove, even eradicate, the plant, but the water hyacinths had turned out to be almost immune to such attempts due to its rapid growth and reproduction. As a second reaction, people started to ask if it could be utilized. In Bengal, East India, Robert Finlow, a government fiber expert, was the first to come up with a working proposal for large-scale water hyacinth utilization. That happened already in 1914 [20].

Initially, the interest in utilization focused on the content of potash and the best way to retrieve that from the plant material. Later, a variety of other uses of the plants have been suggested and tried out.

Already during British colonial times, an intensive debate ranged between the two perceived options—eradication or utilization—both of which in practice turned out to be illusive. In postcolonial East Pakistan and independent Bangladesh, these two lines of approach to the water hyacinths have continued to have their advocates. With time and experience though, the positions have softened and become less antagonistic. "Eradication" has evolved into "control," and most uses have been found to be economic only when the "harvesting"—the physical collection and landing—is paid for from funds for clearing the waterways. These uses today include basket, hat, and furniture production from dried water hyacinth stems;

production of fragrances, perfumes, and colognes from the flowers; treatment of horse skin diseases from the roots; and production of biogas via fermentation of the whole fresh plant.

A special and apparently very successful case is the use of water hyacinths as the bases of floating gardens similar to those in Lake Inle. This has benefitted in particular the poor without their own land but also allows landowners to grow another crop at the time of the year when their land is flooded (more on the uses in the special chapter below).

Banglapedia, the national encyclopedia of Bangladesh, in its 2012 edition, takes a positivistic view of the current situation and credits the work done in the 1930s and 1940s following the Water Hyacinth Act of 1937 with much of the success:

> "By 1947 the scourge of the water hyacinth came under control and in the next decade the rivers of East Bengal became more or less navigable again. Water hyacinth still exists in many parts of Bangladesh, particularly in the haors and beels, but these do not pose any serious problem to navigation or to cultivation any more. The plant is chiefly used as manure; the leaf is used as fodder in floodplains when fodder is scarce. Heaps of water hyacinth, when decayed, make wonderful fertiliser for cropping. Moreover, some crops and vegetables grew luxuriously on the dressed water hyacinth heaps. This has attracted many landless peasants to accumulate and heap water hyacinth to make floating fields for agricultural production."

Case Story: The Nile River

As told above in the section on introduction and spread in Africa and Asia, there are two versions of the story of how and when the water hyacinth reached Egypt and the Nile. One says by the French around 1800 and the other by the British almost a century later. However, the plant was not perceived as much of a problem until the 1960s, when construction of the Aswan High Dam was well under way. This slowed the flow of the Nile River downstreams. Earlier, water hyacinths had been present in irrigation canals and caused some obstruction of water flow and small boat traffic there, but were also seen as a source of fertilizer by the farmers and collected for that use. The main Nile channel was by large free of the plant.

With the slowdown of the overall flow and the stop of the annual flooding, the situation changed. When the water hyacinths were no longer flushed downstream and eventually out at sea, they started to form major clusters occupying also long stretches of the main river. The problem peaked in 1983, when 8400 ha were infested to a level where boat traffic and fishing were impeded [21].

This triggered a response from the government in the form of a large-scale control attempts including the use of herbicides with the aim to keep the main channel fully navigable. The attempts were largely successful, and by 1992 the infestation had been brought down to some 380 ha. Following a stop of the use of herbicides, because of the damage they caused also to crops, the water hyacinths spread again, especially affecting the delta area and the lakes there.

Today, in the lower Nile, water hyacinths are largely controlled by physical means like floating booms to keep them out of certain areas, collection and removal for composting by mechanical harvesting through public works in the main channels, and manual collection by the farmers in the small canals. They are still a problem, but nothing compared to the situation in the 1970s and early 1980s. "They have lost much of their virulence," as one of the high officers at the Ministry of Public Works and Water Resources (MPWWR) puts it.

A main lesson from the history of the water hyacinth in the lower Nile is the role of high water flow and regular flooding. These are factors that prevent the plant from forming thick and wide aggregates, and as long as they prevail, the plant does generally not reach pest status. But without them it might well do so.

In Sudan, above the high dam and Lake Nasser/Lake Nubia, where water hyacinths were first recorded in 1958 [22], the history and the current situation are different.

Also here, however, the exact time of introduction is in question. In a later published article, in a far less prestigious paper [23], the authors claim to have seen and sampled it in 1955–1956. Even with that starting time, though, the further spread was explosive. In the early 1960s, it covered up to 3700 km^2 [24], with a seasonal variation where it was most abundant during the high water season [25].

The associated problems were large. The thick mats, even islands, of water hyacinths, often with other plants including small trees growing on them, interfered with river transportation and increased cost of operation and maintenance of ships and increased travel time. It blocked harbors, irrigation canals, and pumps as well as access to open water for riverside settlers, caused water losses through increased water evaporation, and hindered fishing [26].

To control the pest, a major chemical spraying program using the herbicide 2,4-D was launched. By 1963 the Plant Protection Division of the Sudan Ministry of Agriculture had a fleet of trucks, 42 boats, and three aircrafts engaged for the purpose of securing open water access for commercial boat traffic. A total of some 500 tons were applied annually. The program was successful for a while in obtaining its stated goal but at a high monetary cost and with significant side effects when spray drift cause crop damages.

In 1978 the chemical control program was replaced by one of biological control using the weevil *Neochetina eichhorniae* carried out in collaboration between the Sudanese Council for Research and USDA. The year after, the British Ministry for Overseas Development and the Sudanese government in a joint program release the weevil *N. bruchi* and in 1980 the stem-borrowing moth *Samoedes albiguttalis*.

The biological control program was almost immediately a success. Both short and long time, it reduced the vigor and productivity of the plant, and the population was no more an obstacle to navigation or fishing on the Nile.

In the 1980s the program of culture and release of the biological control agents was terminated both as it was assumed that at least *N. eichhornia* now was firmly established, so there would be no need for further releases, and because of the restarted civil war. Since then there are for the latter reason no comprehensive studies of the water hyacinth situation in Sudan. Dr. Nagat Mubarak El Tayeb in her presentation at the conference of the Standard and Trade Development Facility in 2012 gave a positive picture of "a problem solved." UNEP in its post-conflict environmental assessment of 2007 took a much more guarded approach to the uncertainties.

Case Story: Lake Victoria

Lake Victoria, by surface area, the largest lake in Africa and the second largest in the world, is the scene for one of the most dramatic water hyacinth stories. Lake Victoria is drained by the Nile River, but the water hyacinths didn't reach the lake through that route, instead they came from Uganda via Kagera River. In Lake Victoria they were first spotted in 1989 [27].

By 1992 the plant created local problems preventing fishermen from some villages to reach open waters and clogging some irrigation canals. At the peak in 1997–1998, some 20,000 ha of the lake surface were covered [28], with more or less steady mats along the shores and floating island of up to 100 ha size, which drove with the wind from one part of the lake to another. By that time fishing was severely impaired, and even the largest boats, like train-carrying ferries, could not navigate through the water hyacinth masses. Irrigation canals, pumping stations, harbors, and hydroelectrical plants were unfunctional because of clogging. Furthermore the plants harbored disease-spreading insects like malaria mosquito larvae and snails carrying bilharzia worms.

Infestation of water hyacinths in Lake Victoria in 2006, CSIRO European Laboratory (Photo by Mic Julien)

However, the effects of the water hyacinths were not entirely negative. The ecosystems of Lake Victoria were already subject to several severe stresses. Pollution with sewage and agricultural runoff causing eutrophication was one and probably in itself a reason for the rapid growth and spread of the water hyacinth. The introduction of the Nile perch, a prime food for humans but a vicious predator on the native and often endemic fish fauna, was another and overfishing, especially of the smaller, beach-near living, traditionally caught species, a third. In a way, the water hyacinths helped alleviate those problems, by providing shelter among its root system for the prey of the Nile perch and by reducing fishing pressure. Some endemic cichlids thought to have been exterminated reappeared and even reached some abundance. The low-oxygen zones under the water hyacinth carpets also provided living space for species more tolerant of such conditions like catfishes and lung fish, the four species of which had declined alarmingly during the previous decades [29]. Also plant-eating tilapia increased as function of water hyacinth infestation [30].

All in all, though, the negative effects of the high abundance of water hyacinths far outweighed the positive ones.

After having initiated physical removal with large harvesting barges and shredding machines and considered extensive spraying with the herbicide 2,4-D as a chemical control agent, biological control became the tide changer. *Neochetina* weevils were first introduced to Uganda, Kenya, and Tanzania, the three countries surrounding Lake Victoria, in late 1995, early 1997, and mid-1997, respectively. There was a drop in water hyacinth abundance in the first half of 1998, with a real breaking point in early 1999. By 2001 the water hyacinths were not seen as a problem in Lake Victoria anymore. During the first years of the third millennium, this was hailed as a fantastic success of biological control, although some researchers argued that weather patterns following El Nino also played a decisive role [31]. Others reputed that [32].

In 2003 the biological control project was terminated as it was felt that it was not needed in the lake itself and that it was ineffective in the running waters of the Kagali River, as the weevils washed off the plants when these got temporarily submersed in the current. Instead, in collaboration between the governments of Uganda and Egypt, a mechanical harvesting system was operated near the mouth of the Kagali River. People engaged in the harvesting operation claimed to have removed up to and sometimes over 6000 ton a day [33]. Scientists have tried to estimate the more or less continuous flow of water hyacinths into Lake Victoria via Kagali River in area terms and have come up with figures of some 300 ha per year [34]. Since 2010, when the Uganda/Egypt program ended, this amount of water hyacinths flow unhindered into Lake Victoria. This continues to cause problems, but mostly of a local nature in the vicinity of the inflow, and mechanical and manual removal regularly takes place in key spots like ports. A part of the thus collected plants are used for local small-scale production of, e.g., furniture and baskets.

In December 2006, after extensive media reports about the reoccurrence of massive amounts of water hyacinths in the lake, a survey was done of the situation. It indicated that 36.6 ha of water surface was covered at 16 hot spot sites. In 2010 only a few ha were covered in some basins on the Ugandan side, and in the following year, 2011, only limited reoccurrence of water hyacinth had taken place in the previous hot spot bays of Murchison, Hannington, and Thurston.

In that survey an interesting new feature was also observed in that water hyacinths coexisted with other aquatic macrophytes like *Typha domingensis*, *Azolla*, *Nymphaea*, *Ceratophyllum*, *Najas horrida*, and *Pistia stratiotes*.

It was also observed that water hyacinths now occurred in seasonal succession with elephant grass and papyrus where the latter got established on the water hyacinth mats and gradually took over.

While warning that given the dynamism of water hyacinth, what was observed may turn out to be a temporary situation, and presenting a strategy for future surveillance and control, a guarded positive conclusion, was reached [35].

The summary after the surveys and field visits said that "while there are small patches of water hyacinth across the Lake Victoria Basin, the situation was not out of control."

Control Measures

As noted in connection with some of the case studies above, water hyacinth control options consist of three different types generally referred to as physical, chemical, and biological.

The **physical control** in its simplest form is just people who with minimal equipment such as boots, gloves, and a rake collect the plants and bring them to shore. It's obviously a labor intensive and time-consuming undertaking, where typically hundred of persons could clear a patch of thick water hyacinth growth of the size of 1 ha in a day. There is some risk associated with the activity not only from disease-carrying insects and snails but also from snakes that hide among the plants (therefore the boots and gloves).

Mechanical physical control just means that machines instead of human arms do most of the job. There are many technical models to this, but they fall into two distinct groups: those that collect water hyacinths and those that shred them. In the former case, the collected plant material is generally brought on to land, if it's not used to form floating gardens, where it may be used for a number of different purposes, as described in the next section. In the latter case, the plants are hacked into pieces small enough to have a low chance of survival and regeneration, returned to the body of water from which they came, and left to decompose there.

A large mechanical harvester with a conveyer belt to shore may clear half a ha/day of a dense growth of water hyacinths. A shredder may have the double capacity, but generally speaking, the higher the capacity of the equipment, the larger the risk that pieces of the plant are big enough to be able to regenerate, in which case the whole operation may have removed the plant from one location, but helped them to spread to other places. The cost per ha of physical control with mechanical equipment ranges from $1200 to $2500 with addition to costs for mobilization of equipment. Dependent of the local situation, collected water hyacinths, when landed, might have some (low) value or carry further costs for disposal.

A 90 foot large mechanical harvester (Photo from the University of Florida)

A shredding machine called "Hi-baller" that collects plants, grounds them up, and fires the plant slurry to the shore with built-in water cannons (Photo from the University of Florida)

Chemical control is carried out through spraying a herbicide onto the plants either from trucks or carriages on land, from boats, or from airplanes. The most used chemicals for water hyacinth control are 2,4-D and glyphosate. Costs vary depending on size of treatment area, scale of treatment, and herbicide dosage; costs per ha for materials and application by a contractor are approximately $600 for glyphosate and may range from $1500 to $2500 for 2,4-D [36].

(The original figures are somewhat old and based on US experience, but have been found to hold up fairly well and are often quoted.)

Dependent on the side effects on other organisms in the aquatic system or on crops and other plants on land from wind-driven spray (especially when applied from airplanes), chemical control is generally the least favored control option.

Biological control uses natural "enemies" of the water hyacinth from its native area of distribution. A prerequisite is that they are species specific and don't attack other plants. So far, four more or less suitable organisms have been identified and used for the purpose of water hyacinth control. Two of the agents are weevils of the *Neochetina* family, *N. eichornia* and *N. bruchi* [37], and two are moths, *Niphograpta albiguttalis* and *Agicona infusella* [38].

The adults of both weevils are half a centimeter long and night active. In daytime they mostly hide among leaves near the crone of the plant or in rolled-up leaves. Visually, the difference is not large; *N. eichhornia* has a gray streak in the brown color that *N. bruchi* lacks. Their lifespan is some 9 months, and the females may during that time lay some 400 eggs. There seems to be a difference in the favored place of laying the eggs in that *N. eichornia* mostly does it in the tissue of young leaves, while *N. bruchi* prefers the stem. After short of 2 months, the larvae undergo pupation which takes place on the roots of the plant that is under the water surface. After a week they hatch out as adults.

The adult weevils feed on the epidermis of the leaves forming characteristic scar patterns. The larva bore into leaves and stems, water logging them and opening them up to infections by bacteria and fungi. Typically, weevil-attacked plants float deeper in the water than healthy ones; the production of flowers, leaves, and daughter plants is much reduced; and plant growth is stunted.

Two adult weevils munching on a water hyacinth, International Institute of Tropical Agriculture (Photo by Obinna Ajouno)

The moths have as adults a much shorter lifespan than the weevils. Theirs is counted in days or the odd week. Eggs laid on the water hyacinth leafs in crevices or injures take 4–5 days to hatch. The larvae will initially feed on the surface layer of the leaf, but after a few days, it will borrow into them, buds in particular. While *Niphograpta albiguttalis* predominantly attacks plants on the perimeter of water hyacinth growths, *Agicona infusella* has an affinity for the high and slender specimens that are most found in the center of the plant islands. After 2–4 weeks, the larvae will pupate inside the plant without spinning a cocoon, but with a pre-prepared "window" in the plant tissue for the exit of the imago a week or a little more later. The adult moths are good fliers and can during a night easily cover a distance of up to a kilometer. That is their main advantage over the weevils. Otherwise, they have proven to be less effective. *Niphograpta albiguttalis* has succeeded to establish itself in about half the number to countries to which it has been introduced. *Agicona infusella* is less host specific and has been the least utilized of the four species for water hyacinth control.

Niphograpta albiguttalis

Photo from Wikipedia

Utilization of Water Hyacinths

As has been noted in some of the case studies, water hyacinths are used for several purposes and sometimes are seen as valuable by members of local communities [39]. Mostly, though, these uses are economical only when the costs for collecting the plant and bringing it ashore are paid from a budget for physical control. Among the many proposed and tried out uses [40], the following are those that seems to be most popular:

Production of **yarn** and **rope** for further processing to **baskets**, **hats**, **mats**, and **furniture** [41]. For this, it's the fibers of the stem that's utilized. The stalk is thereby shredded on the length and the fibers are left to dry. The rope production process is then much like that for hemp or jute rope.

Baskets made from water hyacinths, Indonesia

For more up-market uses, like furniture production, the finished rope is treated with sodium metabisulfite ($Na_2S_2O_5$), which, in some applications, is referred to as E223. This functions as a preservative and prevents the rope from rotting.

The rope, of course, can be used as such, but a large number of products, like those mentioned above, can also be manufactured from it.

Rope production from water hyacinths occurs to some extent in most developing countries to which the plant has spread.

Water hyacinth furniture, Indonesia

Production of **paper** and **cardboard**. The quality of paper made from water hyacinth fiber alone is not very good, but when blended to 50% with jute, hemp, or waste paper, it becomes acceptable. The fiber mixture is then heated with calcium and sodium carbonate to produce pulp and often bleaching powder is added. The pulp is dewatered in a hand press and the paper sheets are hung to dry.

Small-scale cottage industry paper making from water hyacinths take place, e.g., in Bangladesh, India, Burma, Vietnam, China, Indonesia, and the Philippines.

In a similar process, **fiber board** can be produced. In this process water hyacinth stalks are chopped, boiled, washed and beaten to pulp, and mixed with waste paper and/or other fibrous plant material as well as china clay after which the pH is balanced. Boards are produced in hand presses and sun dried. The boards are used for indoor partitioning walls and ceiling. Bituminized water hyacinth board has found some use as low-cost roof material.

This type of fiber board production is centered in Bangladesh but also occur in India, in Indonesia, and in the Philippines.

Production of **biogas** from many types of waste and plant material is commonly used all over east Asia [42]. In principle: The organic material is decomposed in an anaerobic process inside an airtight container and methane gas that can be used for cooking, lighting, and heating, and a nutrient-rich slurry, good as fertilizer, becomes the products. Compared to other forms of organic matter, water hyacinths have the drawback of a high water and air content, which lead to low yield in relation to volume and a tendency for the plant material to float on the surface in the container.

It therefore has to be chopped up and beaten before put into the digester. Often, when weather is suitable, it's also left to half-dry before being processed. The best results are obtained when water hyacinths are used as biogas generator feed together with animal wastes, e.g., pig manure [43].

Biogas production with water hyacinths as one of the raw materials is practiced to some extent all over south and Southeast Asia and experimentally in several places in Africa, e.g., Kenya.

More on an experimental scale, charcoal briquettes are also produced from water hyacinths. The pyrolysis (burning with little or no oxygen) that transforms the plant material to charcoal requires relatively dry matter from the start. It has been calculated that to produce 40 tons of briquettes daily, 1300 tons of wet water hyacinths are required, and to dry it an area of 12 ha must be available. In addition outdoor temperature must be high and humidity low.

As said, biogas reactors will produce a nutrient-rich slurry as a by-product that can be used as a **fertilizer**. Foregoing the biogas, nutrient-rich soil can also be produced by aerobic composting. For this minimal equipment is needed. This is the bulk use of harvested water hyacinths, and the organic, nutrient-rich soil that's produced can replace expensive commercial fertilizers.

A version of this is the floating garden, which have become a great success primarily in Lake Inle in Burma and in Bangladesh, as is said in the case studies above. The benefit here, besides the good soil produced by decaying plant material, is the creation of "land" to farm. An easy-to-understand instruction in the making of a floating garden was found in a blog from Lauren Brown [44] and the Borgen Project:

"The following steps are all it takes to make a cheap and sustainable floating garden:

1. Decide on an appropriate size for the floating garden. Generally, rafts are about 8 m long and 2 m wide and are 0.6 m to 1 m deep. The exact size depends on the amount of space and resources available.
2. Collect water hyacinth. This aquatic weed will serve as the base, or raft, for the floating garden. Water hyacinth is fairly abundant in Bangladesh and is free for collection.
3. Lay bamboo poles over the collected plants. The poles should be appropriate to the overall size of the raft.
4. Collect additional water hyacinth and place it on top of the bamboo layer to build the thickness. Weave the water hyacinth into a raft.
5. Once the plants have been woven and the general structure of the raft has been established, remove the bamboo poles.
6. Wait 7 to 10 days and add more water hyacinth to the existing raft.
7. Add a mulch of soil, compost, and cow dung to cover the raft. This layer should total about 25 cm deep. Usually the compost is composed of azola and other easily accessible organic matter.
8. Pick an appropriate place for the raft. Floating gardens should not be placed in waters with tides or currents as the water movement damages the water hyacinth and risks total disintegration of the raft.

9. Plant seeds. The most effective technique is to place a couple of seeds into a ball of compost and tema, an organic fertilizer. These balls are placed in a shaded, protected area while the seeds germinate. Once seedlings sprout, plant them on the raft.
10. Tend the floating garden as appropriate to the crops planted. In Bangladesh the most common crops tend to be leafy vegetables, okra, gourds, eggplant, pumpkin, and onions. Animals like ducks and rodents might be attracted to the rafts. Fencing, even using improvised means like fishing nets, can effectively protect the gardens.
11. Harvest the crops. Rafts can be reused or, if no longer in a useable condition, can be used as compost on a new raft."

Floating gardens under construction in Bangladesh (Photo from the CARE International)

Water hyacinths can also be used as **animal fodder**. It's not rich enough in all that cows, pigs, or chicken require, so it should not be used as a single or main feed but as a supplementary one. It can be, and is often, utilized as an emergency "bridge over" feed during, e.g., dry spells by poor farmers in South and Southeast Asia.

For herbivorous fishes, such as grass carp and tilapia, water hyacinths are adequate food. Experimentally, also manatees and hippopotamuses in captivity have been feed them with some success.

The flowers of the plant is also used in low-volume, high-value applications such as flagrances, perfumes, and certain skin disorder treatments in both humans and animals (horses) [45, 46].

The flowers—and the plant itself—are also used for **decorative purposes**, e.g., in garden ponds, which is an important factor behind the water hyacinth becoming "the worst invasive aquatic plant," as colonialists and other brought them all over the tropical and subtropical world to make their gardens prettier.

A different primary use of water hyacinths, which can easily be combined with all the above and sometimes is, is for **treatment of sewage water**. Domestic sewage with human urine and feces contains a lot of nutrients like phosphate and nitrate. The fast-growing water hyacinths are extremely good at picking up nutrients and thereby clean the water to reduce the risk of eutrophication and algal blooms after final discharge. Their ability to remove substances from water is, however, not limited to nutrients but includes also metals, radioactive compounds, and a variety of organic chemicals that can be found in industrial waste. When the concentration of certain such pollutants gets too high in the water hyacinths, the plant material cannot be used but has to be disposed of as environmentally hazardous goods. It might still, though, be a better option environmentally and economically than to release the waste water untreated or to use physical or chemical water treatment techniques, which also in the end carry sludge disposal problems.

Current Status

The biological control, primarily with the *Neochetina* weevils, has turned the water hyacinth menace into a practically manageable problem, using physical and occasional chemical control as temporary complementary measures. A number of usages of the plant have also become plusses in the socioeconomic balance, even if they (so far) don't outweigh the economic losses in most places.

References

1. Schmitz, D. C., Schardt, J. D., Leslie, A. G., Dray, F. A., Osborne, J. A., & Nelson, B. V. (1993). The ecological impact and management history of three invasive alien aquatic plants in Florida. In B. N. McKnight (Ed.), *Biological pollution—The control and impact of invasive exotic species* (261 pp). Indianapolis: Indiana Academy of Science.
2. Muschler, W. T. (1912). *A manual on the flora of Egypt*. Berlin, Germany: R. Friedlander und Sohn. 249 pp.
3. Simpson, N. D. (1932). *A report on the weed flora of irrigation canals in Egypt* (124 pp). Cairo, Egypt: Ministry of Public Works and Irrigation, Government Press.
4. Stent, S. M. (1913). *Water hyacinth* (Publication No. 68). Union of South Africa: Dept. of Agriculture.
5. Edwards, D., & Munsil, C. F. (1975). *Eichhornia crassipes* in South Africa—A general review. *J. Limnol. Soc. Sth. Afr., 13*, 16–17.
6. Pooley, E. (1998). *A field guide to the wild flowers of KwaZulu-Natal and the Eastern Region* (pp. 448–449). Durban: Natal Flora Publications Trust.
7. McLean, K. Water hyacinth as a serious pest in Bengal. *Agricultural Journal of India, 17*, 23–40.
8. Rao, V. S. (1988). *Principles of weed science* (544 pp). New Delhi, India: Oxford and IBH Publishing.

9. Molle, F., & Keawkulaya, J. (1998). Water management and agricultural change: A case study in the upper Chao Phraya Delta. *Southeast Asian Studies, 36*(1), 32–58.
10. The Chao Phya in Transition. Steve Van Beek (p. 168).
11. Pieterse, A. H. (1997). *Eichhornia crassipes* (Martius) Solms. In I. Faridah Hanum & L. J. G. van der Maesen (Eds.), *Plant resources of South-East Asia No. 11: Auxiliary plants* (pp. 118–121). Leiden, The Netherlands: Backhuys Publisher.
12. Teygeler, R. (2000). Waterhyacintpapier. Bijdrage aan een duurzame toekomst/Water hyacinth paper. Contribution to a sustainable future [bi-lingual]. In Torley and Gentenaar (Eds.), *Papier en Water/Paper and Water* (pp. 168–188). Rijswijk: Gentenaar & Torley Publishers.
13. Kirunda, J. Human Flower Project, Water Hyacinth—Double Edged.
14. Gopal, B., & Plants, A. (1987). *Studies 1: Water hyacinth* (p. 5/43). London: Elsevier.
15. Gopal, B., & Sharmer, K. (1981). *Water hyacinth (Eichhornia crassipes), the most troublesome weed of the world.* New Delhi: Hindasia.
16. Bosch, J. H. *The Origin and Spread of Waterhyacinth, Eichhornia crassipes.* (Mart.) Solms.
17. Annual Meeting Weed Science Society of America 69 Abstract. (1978). In Allan, M. (Ed.), *Weeds: The unbidden guests in our gardens* (p. 25). New York: The Viking Press; Whittle, T. (Ed.), *The plant hunters* (pp. 25–7).
18. The Water Hyacinth Act of 1917 (repealed by SLORC Law No. 1/92). *The Burma Code.* Vol. IV, pp. 296–297.
19. Sidle, R. C., Ziegler, A. D., & Vogler, J. B. (2007). Contemporary changes in open surface area of lake Inle, Myanmar. *Sustainability Science, 2*, 55–65.
20. Iqbal, I. (2009). Fighting with a weed: Water hyacinth and the state in Colonial Bengal, c. 1910–1947. *Environment and History, 15*(1), 35–59.
21. Navarro, L. A., & Phiri, G. (Eds.). (2000). Chapter 9: Water hyacinth in Africa and the Middle East: A survey of problems and solutions.
22. Gay, P. A. (1958). Eichornia crassipes in the Nile of Sudan. *Nature, 182*, 528.
23. Heinen, E. T., & Achmed, S. H. (1964). *Water hyacinth control on the Nile river, Sudan.* Khartoum: Information Production Centre, Department of Agriculture.
24. Mohammed, B. F. *The ecology of waterhyacinth in the White Nile, Sudan.*
25. El Tayeb, N. M.. *Successful biological control of water hyacinth (Eichhornia crassipes) by Neochetina weevils in Sudan.* Presentation given at the STDF Conference on International Trade and Invasive Alien Species, 12–13 July 2012.
26. Obeid, M. (Ed.). (1975). *Aquatic weed in the Sudan with special reference to water hyacinth* (150 p). Khartoum, Sudan: National Council for Research.
27. Twongo, T. (1991). Status of the water hyacinth in Uganda. In A. Greathead & P. J. de Groot (Eds.), *Control of Africa's floating water weeds* (pp. 55–57). Zimbabwe: CAB International.
28. Albright, T. P., Moorhouse, T. G., & McNabb, J. (2004). The rise and fall of water hyacinth in Lake Victoria and the Kagera River Basin, 1989–2001. *Journal of Aquatic Plant Management, 42*, 73–84.
29. Goudswaard, Kees (P. C.), Witte, F., & Chapman, L. J. (2002). Decline of the African lungfish (Protopterus aethiopicus) in Lake Victoria (East Africa). *African Journal of Ecology 40*(1):42–52.
30. Njiru, M., Othina, A. N., & Wakwabi, E. *Impact of water hyacinth, Eichhornia crassipes, on the fishery of Lake Victoria, Kenya.* Report of phase 1 from The Lake Victoria Environmental Management Project (2012) Lake Victoria Basin Commission.
31. Williams, A. E., Duthie, H. C., & Hecky, R. E. (2005). Water hyacinth in Lake Victoria: Why did it vanish so quickly and will it return? *Aquatic Botany, 81*, 300–314.
32. Wilson, J. R. U., Ajuonu, O., Center, T. D., Hill, M. P., Julien, M. H., Katagira, F. F., Neuenschwander, P., Njoka, S. W., Ogwang, J., Reeder, R. H., & Van, T. (2007). The decline of water hyacinth on Lake Victoria was due to biological control by *Neochetina spp. Aquatic Botany, 87*, 90–93.
33. Retrieved from http://www.newvision.co.ug/news/641529-water-hyacinth-threatens-to-choke-lake-victoria-again.html

34. Twongo, T., & Balirwa, J. S. (1995). *The water hyacinth problem and the biological control option in the highland lake region of the Upper Nile basin—Uganda's experience*. Paper Presented at the Nile 2000 Conference: Comprehensive Water Resources Development of the Nile Basin: Taking Off, 13–17 February 1995, Arusha, Tanzania.
35. *Lake Victoria Basin Water Hyacinth Surveillance, Monitoring and Control Strategy, 2012–2030*. A report from the Lake Victoria Basin Commission with support of the World Bank and SIDA.
36. Gibbons Jr., H. L., & Sytsma, M. D. (1999). *Guide for developing integrated aquatic vegetation management in Oregon*. Portland, OR: Center for Lakes and Reservoirs, Portland State University.
37. Center, T. D., Dray, F. A., Jubinsky, G. P., & Leslie, A. J. (1999). Water hyacinth weevils (Neochetina eichhorniae and N. bruchi) inhibit water hyacinth (Eichhornia crassipes) colony development. *Biological Control, 15*, 39–50.
38. Julien, M. H., Griffiths, M. W., & Stanley, J. N. (2001). *Biological control of water hyacinth. The moths Niphograpta albiguttalis and Xubida infusellus: Biologies, host ranges, and rearing, releasing and monitoring techniques*. ACIAR Monograph No. 79, 91p.
39. Herfjord, T., Osthagen, H. & Saelthun, N.R. (1994). *The water hyacinth* (39p.). Oslo, Norwegian: Agency for Development Cooperation. Herklots, G.A.C.
40. Lindsey, K., & Hirt, H. M. (2000). *Use water hyacinth!: A practical handbook of the uses of water hyacinth from across the world*. Winnenden: Anamed.
41. (1976). *Making aquatic weeds useful: Some perspectives for developing countries*. Washington, DC: National Academy of Sciences.
42. Duke J. *Eichhornia crassipes in handbook of energy crops*. Available only online.
43. Eden, R. (1994). *Water hyacinth utilisation*. Thesis, Warwick University, UK.
44. Retrieved from http://borgenproject.org/how-to-make-a-floating-garden/
45. Duke, J. A., & Wain, K. K. (1981). *Medicinal plants of the world. Computer index with more than 85,000 entries* (Vols. 3).
46. Oudhia, P. (2001). Traditional medicinal knowledge about a noxious weed, Jal kumbhi (*Eichhornia crassipes*), in Chhattisgarh (India). *Aquaphyte 21*(2).

Starlings in North America

> *"Without seeming to care whether the benefiting farmer thanks him or reviles him, he hurries with jerky steps about the farms and gardens in the summer time, carring more than 100 loads of destructive insects per day to his screaming offspring, cramming his own stomach full of such foods as Japanese beetles, caterpillars, and cutworms. With complete indifference to angry protests, he finds roosting places in warm cities in the winter, going out each morning, a faithful commuter in reverse, to earn his bread in the surrounding countryside."*
>
> *(Rachel Carson)*

A Little About the Bird Itself

The starling belongs to a fairly large family of birds whose natural distribution covers most of the old world. Also the species itself, which Linné gave the name "vulgaris," the common one, has a wide natural distribution in Europe, North Africa, the Mideast, Iraq, Iran, Pakistan, North India, Nepal, and Western and Northern China.

The bird is about 20 cm, has typically a wing span of just under 40 cm, and weighs 60–100 g. The basic color is black with iridescent blue, purple, or green spots. After molting in late summer, it appears white spotted as the new feathers have a white tip. As this during and after winter is worn down, the white spots gradually disappear. The legs are pink to gray red, the beak is brown black in winter and yellow in summer, that of the male more bright than the female's. Young birds are dark brown.

A. Jernelöv, *The Long-Term Fate of Invasive Species*,
DOI 10.1007/978-3-319-55396-2_10

Starling (*Sturnus vulgaris*) (From www.allaboutbirds.com)

On the ground it moves by walking or running, not jumping. It flies with rapid wing movements and maneuvers with great skill, which is particularly obvious when it moves in large flocks that whirl away, fractions up into smaller units, and reunites in dizzying speed. The flocks continuously regroup, lack visible leaders, with every bird navigating in relation to the movements of its closest companions and based on a common instinctive feeling for general directions and what constitute good and bad landing sites. During migration these flocks can be very large, sometimes consisting of up to a million birds.

The starling nests in cavities, where the male constructs a nest made of twigs and grass, feathered with dry leaves and feathers and decorated with flowers and green stuff (!) to attract the female. (When she has accepted the site and the partner, they often rebuild the nest on her initiative.) Often, many pairs nest close together, thereby forming loose colonies. The female lays four to five blue eggs, which the parents take turn to sit on, but mostly with longer passes for the female. The young ones are diligently fed by both parents, who also remove feces and litter, and are fledged in 3 weeks. In most areas, a couple can raise 2 or 3 L in a breeding season.

The males are not fully monogamous but will, if given the opportunity, mate with several females. Occasionally, they will also build more than one nest and participate in the incubation and feeding of several litters in parallel. More common is that the female who is fertilized by "married" male, in a cuckoo-bird fashion, will lay her eggs in the nest of the other female. Cases have also been observed where several females and one male jointly incubate and feed a large bunch of offspring.

During the time spent at the nesting site, the male often sings, especially at dawn and dusk. The song is often referred to as a whistle, but one with many variations and individual characteristics. The male starling can also ape the songs of other birds and sounds, which has encouraged many to try to teach caged starlings to talk like parrots.

The starling mainly eats insects and worms, but at places and times of the year, when such feed is difficult to get by, they also consume fruits, berries, and seed. Of animal feed, starlings consume 10–20 g daily and about twice that amount if they are limited to a vegetable diet.

They can be aggressive, particularly in nesting site competitions, and drive away other species.

A unique feature, anatomically, is the muscle setup around the beak. The muscles that open it are very well developed. This in combination with the setting of the eyes, which make the bird able to see along the whole beak and allow it to widen a hole in the ground or a rotten tree stump, where it felt or sighted something of interest. This facilitates the catch of, e.g., wood-living larvae, millipedes, and worms.

A swirling flock of starlings at the time of migration (www.telegraph.co.uk)

Introduction to North America and the Subsequent Spread

The first attempt to introduce starlings to North America took place in Westchester, Pennsylvania, in 1850, according to available sources, and was followed by another attempt in Cincinnati, Ohio, in the winter 1872–1873 [1]. Neither of these appear to have been successful, as nothing more is said about these starlings. The first attempt to introduce them, which was not an apparent failure, took place in May 1889 in Portland, Oregon, where 20 pairs were released. In 1901, it was noted that a couple of pairs nested there, but since then there are no more reports of that population [2]. The origins of the starlings that live in North America today are in all likelihood two releases, each of 80 starlings, which took place in Central Park in New York City in April 1890 and May 1891 [3]. Behind the successful attempt was the American

Acclimatization Society, an association that had taken upon itself to enrich the fauna and flora of America with useful and beautiful species from other continents, Europe foremost. In 1877, a pharmacist by the name of Eugene Schieffelin became the chairman of the group. He had a burning interest in Shakespeare, besides ornithology, and was determined to introduce all bird species that the admired author had mentioned in his works. The starling appeared in the first part of Henry IV, where its ability to speak (?) gives the bird a role in a plot against the king.

The first place where a successful starling nesting took place can be seen as symbolic: under the roof of the American Museum of Natural History. Less than a decade later, the starlings were firmly established in New York City with surroundings but hadn't spread further than that. After that the spreading picked up pace, and the starlings reached Hartford having followed Connecticut River upstream and Newburgh along the Hudson River by 1905. South- and westward they had gotten to Princeton, New Jersey. A lone pair nested in Trevose, Pennsylvania, and wandering winter flocks reached Philadelphia. Five years later, they were established in and around the city and spread out in all directions, much the same as they had done from New York earlier.

Ithaca was reached in 1916, which is also the first year in which they were seen nesting in Washington, DC, where flocks had spent the winter the year before. The border to Canada was crossed in the summer of 1919, and 3 years later, they were sighted in Ottawa and nested in Burlington on the western shore of Lake Ontario. The year after, they were also present in Port Dover at Lake Erie and reached Quebec City and Montreal. In 1927 they could be found in numbers over large part of the provinces of Quebec and Ontario. By then they were also established around Lake Michigan and in some places in Indiana; eastern Kentucky; eastern Tennessee; Athens, Georgia; and northern South Carolina. The year after the starling's distribution ranged from Maine to the Mississippi Valley and from South Canada to the Mexican Gulf. In the north they were summer-nesting birds and in the far south present mostly over winter [3].

In 1940 it was estimated that the number of starlings in North America had reached 50 million and that they were spread over a surface area of 7 million km^2 [4].

By the end of 1940, the starlings had reached Salt Lake City in Utah, and by the early 1950s, they had passed the Rocky Mountains in their westward spread and reached, among other places, Williams Lake in British Colombia, Roxboro in Washington State, and La Grande in Oregon [5].

What was left then was California. Single starlings had arrived there in the early 1950s, and in the first part of the decade, they consolidated their presence in the northern part of the state. The Audubon Christmas Bird Count shows a strong increase of the numbers and an expansion of the colonized territory there from the middle 1950s, and the trend continued throughout the 1960s.

In Southern California, the process went at a slower pace. The starlings arrived there in the late 1950s, but not till the mid-1960s the expansion steamed up. In 1971, there was an unexplained backlash according to the bird counts [6].

It took till late 1970 until the starlings were fully established in Southern California and on the Northern Mexican west coast and thereby had reached the full current distribution in North America. The total number of starlings were then estimated to be over 200 million [7, 8].

Since about 1970, the starling territory in North America has not expanded. The area with the highest density (= numbers per surface area) had moved west from New York and New England and can now be found in the three Cs triangle (Chicago, Cleveland, and Cincinnati). The number of starlings is still often given as 200 million, despite the fact that the North American Breeding Bird Survey shows a slow but very steady decrease over the last four decades with an average yearly decline of just over 1% [9].

The accumulated effect thereof would mean that the number of starlings today would be around 140 million. Some organizations like Internet Center for Wildlife Damage Control and Extension and America's Science Learning Network also use this lower figure.

Ecological Effects

The real majority of the articles that deal with the ecological effects of the starlings talk about that the aggressive birds drive away native cavity-nesting birds from their nesting sites and document that this has reduced the local population of some desired species. The assumption often is that this has damaged the effected species also on a national level. More than a 1000 articles on this theme dealing with almost 100 native species can be found by searching a suitable data base. Few attempts have been made to study the question of starling interaction with native birds on an overreaching level. In one such study based on Christmas Bird Counts and Breeding Bird Survey, Koenig found a negligible overall effect of starlings on 27 species of cavity-nesting birds that most often are those seen as disadvantaged from the competition for nesting sites. For a small group of woodpeckers, the sapsuckers, (*Sphyrapicus*), there was a negative geographic and time correlation in occurrence that indicated that the decreased numbers of these birds were caused by the starlings [10].

Economic Effects

Also the reported economic damages almost always relate to local effects, even the most often quoted figure damage to agriculture on the national US level [11]. Pimentel's figure from year 2000 of $800 million is based on an estimated average damage figure of $5 per ha.

At an international symposium on the theme "Managing Vertebrate Invasive Species" in Fort Collins in 2007, Linz presented a number of damage occasions [12]. For example, three beef cattle breeders in Kansas reported in 1999 damage caused by the birds to an amount of $600,000; estimates from Colorado in 1968 said that overwintering starlings consumed beef cow feed for $84 per 1000 birds (much more with current feed prices). In Idaho, ranchers counted with a daily consumption of beef cow feed by starlings of 15–20 tons. Damages on growths of grapes and cherries in millions of USD were also reported.

All in all, however, Linz and coworkers believe that the damages caused by the spread of diseases by the starlings are even more economically damaging than those caused by their feed consumption. They exemplify cases of disease outbreaks among domestic animals, each with a price tag of over a hundred thousand dollars. (The weakness of the argument is possibly the connection to starlings. Even if the birds have been identified as carriers of pathogens in question in unrelated studies, it's not proven that they were the culprits in the cases where the economic damage was assessed.)

The starlings also cause economic damage to the aviation industry through collision with airplanes (seldom fatal for passengers and crew) and cost to society for cleaning streets and buildings when major flocks have spent the night in cities. Some examples of these costs were also reported.

Practically all reports on economic effects of starlings focus exclusively on the damage side. At the occasions when benefits are mentioned, no attempts are made to quantify them. Starlings prefer an animal diet of insects and worms and during spring and summer that's basically what they eat. Hundred and forty million starlings consume over 2000 tons of such organisms a day and mostly in agricultural environments. Ecologically as well as economically, this ought to be significant also on a continent the size of North America, and for agriculture, the economic effect is likely to be positive. Back in 1921, this argument was made in a report from the US Department of Agriculture [13], but later the agricultural lobby there and elsewhere has totally focused on damages and damage compensation.

Control Efforts

During close to a century, sustained efforts have been made to control the starling population. The methods range from shooting, netting, and nest destruction to poisoning.

During the last 50 years, some 100 million starlings have been killed by governmental agents. In 2008, the number was 1.7 million [14].

Public Perception and Official Policies

The starlings have been cursed as an occupant of nesting sites rightfully belonging to native and more appreciated birds, called a plaque, a menace to the economy of hardworking, honest farmers and an agent spreading diseases and dirt, but the accusations have never become as rancorous and spiteful as those against the house sparrows were at the height of the "sparrow war."

While the establishment closed ranks against the sparrows, the starlings have had their prominent defenders, among them Rachel Carson, who later achieved world fame for her book *Silent Spring*. In an article with the title *How About Citizenship Papers for the Starling?* in the June–July 1939 issue of *Nature* magazine, she wrote:

> "In spite of his remarkable success as a pioneer, the starling probably has fewer friends than almost any other creature that wears feathers. That fact, however, seems to be of very little importance to this cheerful bird with glossy plumage and stumpy tail. Without seeming to care whether the benefiting farmer thanks him or reviles him, he hurries with jerky steps about the farms and gardens in the summer time, carrying more than 100 loads of destructive insects per day to his screaming offspring, cramming his own stomach full of such foods as Japanese beetles, caterpillars, and cutworms. With complete indifference to angry protests, he finds roosting places in warm cities in the winter, going out each morning, a faithful commuter in reverse, to earn his bread in the surrounding countryside.
>
> On one point ornithologists are pretty well agreed--the starling is here to stay. Shall we then continue to regard him as an alien or shall we conclude that his successful pioneering and his service in insect destruction entitle him to American citizenship?
>
> As to economic worth, a pretty complete auditing of the starling's books has already been performed by the Department of Agriculture." [15]

A flock of starlings in flight, for a moment forming the image of a gigantic bird (From webodysseum.com)

Rachel Carson doesn't give any reference to "the complete auditing of the starling's books by the Department of Agriculture," but presumably it's the article by Kalmbach and Gabrielson from 1921 [13] she refers to.

She was not the only celebrity who came to the defense of the starlings, not by far. Most of them were in all likelihood genuine, but lately a somewhat comic mistake of identities helped spread the message.

It all started with an article in *The New York Times* in 2007 [16], with the title *Flight Patterns*. It was a somewhat philosophic piece by Jonathan Rosen, who in connection with the article was presented as "the editorial director of Nextbook." It was also said that his book about bird watching, *The Life of the Skies*, would be published next year.

In the article was the line "It isn't their fault that they treated an open continent much as we ourselves did." The article might have been all but forgotten had it not been picked up by Steve Mirsky, who in *Scientific American* in 2008 [17] wrote "As bird-watcher Jeffrey Rosen put it in a 2007 New York Times article," followed by the quotation above.

Jonathan had become Jeffrey and Jeffrey Rosen is a celebrity. He's a law school professor at George Washington University in DC, a regular contributor of articles to *The New York Times*, and has been called "the nation's most widely read and influential legal commentator." With his name attached to it, and the honorary title of "bird watcher" not forgotten, the message became viral on the web, especially after 2010, when it became common to attach it to images of swirling starling flocks on YouTube and other internet media.

Several organizations with a more tolerant attitude to starlings have also appeared on the scene, e.g., Compassionate Action Institute and Starling Talk, who also care for and try to rehabilitate injured and orphaned starlings.

While there nowadays is a more "forgiving" and accepting public attitude to this bird, with lots of admiration expressed for their flock flights, the official harshness remains. Among vertebrates, rats are the only organism on which more money is spent on extermination.

The Role of Diseases and Parasites

As in the case of other invasive species, the role of diseases and parasites for the invasive success is discussed also with regard to the starlings. Arguments are made for two opposing models. The argument goes that the immigrant birds are successful as they left (some of) their tormentors behind. In the case of starlings, that school can find support in a study of intestinal parasites with starlings in Europe and North America [18]. According to it, there were 26 species in the group Trematoda, 12 in Cestoda, 26 in Nematoda, and six in Acanthocephala in Europe, while there in North America, in the groups, were only four, five, ten, and three species, respectively. Present in both populations were two, three, six, and zero species of parasitic intestinal worms.

The other school has it that invasive species carry pathogens, to which they themselves have a certain degree of tolerance, and that they infect and weaken their more sensitive native competitors, thereby creating an advantage for themselves. Also for this view, some support can be found in certain investigations [12], where North American starlings had been found to carry bacteria, viruses, fungi, and protozoa that give rise to the following diseases: erysipeloid, salmonellosis, pasteurellosis, avian tuberculosis, streptococcosis, yersiniosis, vibriosis, listeriosis, meningitis, encephalitis, aspergillosis, blastomycosis, candidiasis, cryptococcosis, coccidiosis, American trypanosomiasis, toxoplasmosis, chlamydiosis, and "Q-fever," which is spread with avian lice and ticks. (Many of these can also infect man and other mammalians.)

Another Possibly Contributing Factor to the Invasive Success

A large number of attempts have been made to explain why certain species have become successful invaders, while others, who were also introduced, failed to establish themselves. Also the starlings in North America have had their share of analysis and speculations. However, one possibly contributing factor to their invasive success, which has attracted little or no attention, is that they were preceded as successful invaders by one of their favorite food items—earthworms. Unintentionally introduced from Europe by the early settlers, they colonized the continent ahead of the starlings and constituted a readily available food source, when the birds arrived. In Europe, the starlings happily consume earthworms, although there they have competitors who zoom in on the same prey. In North American there was much less competition for this food source. As earthworms have largely been absent at least since the last glacial period, birds and mammalians there, in their later evolution, have not specialized in catching these crawlers. Thus, when the starlings arrived, there was an underutilized feed source available that they were accustomed to use.

How Do the Starlings Fare in Their Native Europe?

In Great Britain, the number of starlings has decreased during the last several decades. BTO, the British Trust of Ornithology, which has been counting them for a long time, reports that their numbers have dropped by two thirds since the 1970s. The decline is confirmed by data from UK Breeding Bird Survey. The drop has been particularly strong in England, more than 80%, while the starling population in, e.g., Northern Ireland, actually increased [14]. In England and Wales, the 1-year survival rate has dropped from 30 to 15%, most likely dependent on lack of feed, when cattle-grazing grounds have been reduced and thereby available insects. As a result of the sharp drop in numbers, the starling is now red-listed in UK as a species with a threatened future.

Also in many other parts of Northern Europe, the number of starlings has dropped. The largest population declines since the 1980s have been in Sweden, Finland, the Baltic states, and Russian Karelia, while the decreases have been milder in Poland, Germany, Hungary, Austria, and Switzerland. In south of Europe, picture is the opposite, with strong increases in starling numbers in Italy and south of France [19]. In Northeast Spain, where starlings used to spend only winters, they started nesting in the 1960s, and the local population is growing, but so far without much expansion of the territory.

Even to Iceland, the starlings have spread, and they nest there since the 1940s, in slowly growing numbers.

Summary

From the introduction in 1890, it took about 80 years for the starlings to spread over the parts of the North American continent with a, for them, suitable climate. Their number at that time, in 1970, was estimated to 200 million. Since then, the population has slowly but steadily declined, on an average by just over 1% a year, and the current count is about 140 million.

The American authorities still regard it as a pest and devote comparatively large resources to combat them.

References

1. Phillips, J. C. (1928). *Wild birds introduced or transplanted in North America*. U.S. Department of Agriculture Technical Bulletin No. 61, 64p.
2. Lord, W. R. (1902). *A first book upon the birds of Oregon and Washington* (p. 304). Oregon, Portland: W. R. Lord.
3. Cooke, M. T. (1928). *The Spread of the European Starling in North America*. United States Department of Agriculture, Circular No. 40.
4. Wing, L. (1943). Spread of the starling and English sparrow. *Auk, 60*, 74–87.
5. Kessel, B. (1953). Distribution and migration of the European Starling in North America. *The Condor, 55*(2), 49–67.
6. DeHaven, R. W. (1973). Winter population trends of the starling in California a difference in status between Northern and Southern California is discerned. *American Birds, 27*(5).
7. Feare, C. J. (1984). *The starling*. New York: Oxford University Press.
8. Ehrlich, P., Dobkin, D. S., & Wheye, D. (1988). *Birder's handbook: A field guide to the natural history of North American birds* (p. 785). New York: Touchstone.
9. Sauer, J. R., Hines, J. E., Fallon, J. E., Pardieck, K. L., Ziolkowski Jr., D. J., & Link, W. A. (2012). *The North American breeding bird survey, results and analysis 1966–2011*. Laurel, MD: USGS Patuxent Wildlife Research Center.
10. Koenig, W. D. (2003). European starlings and their effect on native cavity-nesting birds. *Conservation Biology, 17*, 1134–1140.
11. Pimentel, D., Lach, L., Zoniga, R., & Morrison, D. (2000). Environmental and economic costs of nonindigenous species in the United States. *Bioscience, 50*, 53–65.

12. Linz, G. M., Homan, H. J., Gaukler, S. M., Penry, L. B., & Bleier, W. J. (2007). European starlings: A review of an invasive species with far-reaching impacts. In G. W. Witmer, W. C. Pitt, & K. A. Fagerstone (Eds.), *Managing Vertebrate Invasive Species: Proceedings of an International Symposium.* USDA/APHIS/WS.
13. Kalmbach, E. R., & Gabrielson, I. N. (1921). *Economic value of starling i United States.* U.S. Department of Agriculture, Bulletin No. 868.
14. Avian Web. *European starling and common starling*. Retrieved from http://www.avianweb.com/europeanstarlings
15. Carson, R. (1939). How about citizenship papers for the starlings? *Nature Magazine, 32*, 317–319.
16. Rosen, J. (2007). Flight pattern. *New York Times.*
17. Mirsky, S. (2008). Shakespeare to blame for the introduction of European starling to U.S. *Scientific American.*
18. Hair, J. D., & Forrester, D. J. (1970). The helminth parasites of the starling (Sturnus vulgaris L.): A checklist and analysis. *The American Midland Naturalist, 83*, 555–564.
19. Pasquali, R. (1984). Resident populations of starlings *Sturnus vulgaris* in southern and Central Italy. *Rivista Italiana Di Ornitologia, 4*, 221–229.

The Argentine Ant on Mediterranean Shores

The Ant Itself

The Argentine ant (*Linepithema humile*) is a dark, red-brown ant, only about 3 mm long with queens that can measure up to a cm. It is native to the Rio de la Plata river drainage basin in Northern Argentina, Uruguay, Paraguay, Bolivia, and Southern Brazil, where it lives in small, inconspicuous colonies like so many other ant species. It has become a major invasive species in many parts of the world where the climate is of Mediterranean type, first and foremost on the northern shores of the Western Mediterranean itself. The basis for its invasiveness is a tendency to form supercolonies, linked anthills with numerous queens, and un-antagonistic collaborative behavior between billions of members, even when they are unrelated.

A. Jernelöv, *The Long-Term Fate of Invasive Species*,
DOI 10.1007/978-3-319-55396-2_11

An Argentine ant queen and worker (Photo from BBC Earth News)

As with many other ant species, the Argentine ant will consume a variety of feed—from other insects and worms, frogs, lizards, small birds, and mammalians (if they find the latter dead or can overrun them in very large numbers), to sweet things such as nectar and the excretion of aphids. Human food and pet feed are delicacies from the ants' perspective.

The eggs the queen lays hatch after a month, and after two more months the larvae will pupae. Another two weeks later, after metamorphosis, they hatch as adults. Argentine ant nests have many queens, and colonies spread through a process called "budding," in which a queen simply walks away with her entourage and builds a new nest not far from the first. This is a slow process and on its own, through budding, a colony might advance the modest distance of 150 m in a year.

The Argentine ant doesn't sting or bite humans. In that way (only) is it benign.

Introductions to the Mediterranean Area

In the days when sailing ships carried the cross-Atlantic trade, the Portuguese islands of Madeira were important stopovers where food and water could be replenished. This was especially so for the vessels sailing between Brazil and Portugal. At Madeira, the first specimen of *Linepithema humile* was collected sometime between 1847 and 1858 [1, 2], well before the type specimen was collected in Argentina in 1866 [3]. As the presence of this particular ant was not registered in the collection until much later, 1882 is mostly given as the year in which the Argentine ant was first found on Madeira [4, 5].

Given the trade routes along which most commerce traveled in those days, it is not surprising that Lisbon was the next place close to the Mediterranean where the ant was found, in 1900 [6]. In fact, all of the earliest recorded sightings from continental Europe were from Portugal [2, 7, 8]. Lisbon was followed by such places as Genoa, Italy (1902) [9], Marseilles, France (1905 [10] or 1906) [11], Northwest Spain (1916) [12], Monaco (1921) [13], Southeast Spain (1923) [14], and Sicily, Italy (1926) [15]. Bosnia and Algeria also reported the presence of the Argentine ant, in 1916 and 1923, respectively [2].

Zoologists noted and reported the occurrence of the ant did not, by and large, mean that its presence out-of-doors initially raised any specific concerns, although an exception is an article by Martins that appeared in 1907 and already in the title referred to "a terribly invasive ant." The presence of the Argentine ant indoors, however, did cause alarm. Beginning in Northern Ireland in 1899 [16], horror stories were told and retold with ever more lurid details about millions of small ants living under floors, in walls, cupboards, or ceilings that were capable of finding any trace of food. Kittens were said to have been suffocated to death by them, and fears were spread that babies might be at risk of the same fate. As reports of the ants being found indoors started coming in from Germany, Belgium, Scotland, Poland, England, Czechoslovakia, Austria, Switzerland, Sweden, the Netherlands, and Norway, those horror stories followed them.

During much of this time, the Argentine ant population on the Iberian shores of the Atlantic Ocean and the Mediterranean Sea developed in peace, budding off colonies and spreading slowly. But with its ability to hitchhike on man's vehicles, the ant also took numerous large and small leaps that hastened its expansion.

The discovery in the mid-1990s of a supercolony of Argentine ants on the Ishikari coast of Hokkaido, Japan, was a sensation and triggered around-the-world attention. That colony spanned a shoreline distance of almost 3 km, consisted of 45,000 nests, more than 300 million workers, and an estimated one million queens [17].

When researchers hurried to check other places that *L. humile* had invaded, looking for other supercolonies, they were richly rewarded. In the USA, one of the colonies found there was aptly named "California Large." It stretches for almost 1000 km along the Pacific coast from the Mexican border to San Francisco. Somewhat later a supercolony was found in Melbourne, Australia, but the largest one was found in Europe. On the Atlantic and Mediterranean shores, from Northwest Spain to northern Italy, along a stretch of 6000 km, the Argentine ant had formed the largest recorded society of multicellular organisms. The number of queens in that supercolony can in all likelihood be counted in the tens of millions and the workers in the many billions [18].

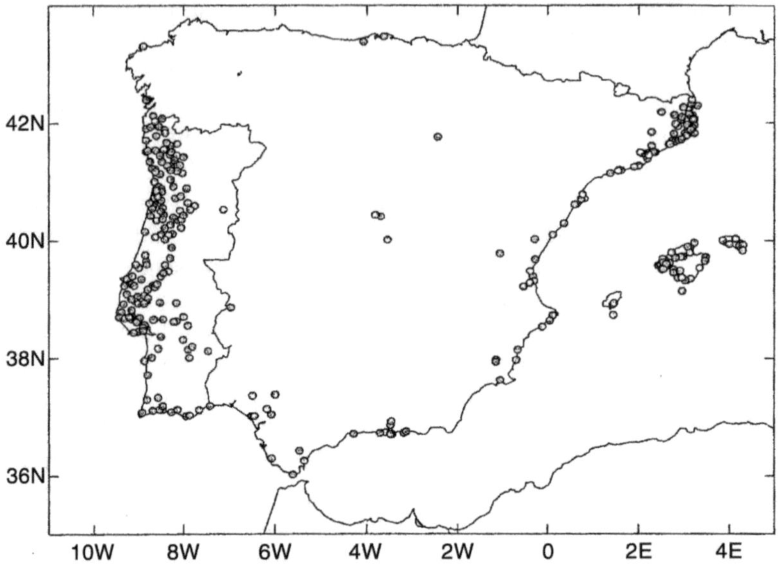

Map over the distribution of *Linepithema humile* in southwestern Europe (From Geraud et al. [18])

Supercolonies: How and Why?

The type of ant colony that forms the mental archetype of most students consists of one female, fertilized in-flight, and her offspring—a large number of hardworking, sterile, haploid females constantly engaged in maintaining the colony, and a relatively few diploid winged males and females, hanging around waiting for their time in the air and the sexual encounters they will have there. They share a genetic setup and a scent and are nonaggressive toward one another. When encountering ants from other colonies with different scents and marker genes, however, they become aggressive and often fight to death.

The ants in supercolonies are different. For a start, they mostly do not have the in-flight mating system but mate in the nest, which in a small colony means with their siblings. Such a colony will, however, if it prospers, soon have a large number of egg-laying queens. Over time, they will be more and more distantly related. For this to function, the ants in supercolonies have to be less discriminating and accept (or not notice) a larger variation in scent of their colony coinhabitants.

Not all fertilized females remain in the nest. They cannot fly, so they just walk out with their entourage and form a new nest in the near vicinity. The two nests, and the many more that will be formed in the same way, still belong to the same colony.

Walking is, however, not the only way the ants can spread. The Argentine ant, in its native range along Rio Parana and other tributaries to Rio Plata, has a distribution linked to the watercourses, which suggests a waterborne spreading mechanism. As an invasive species, the ant undoubtedly has hitchhiked on human vessels and vehicles. How, when that has happened and a queen and some worker ants have formed a new nest perhaps tens of kilometers or even an ocean away from the rest of their colony, and they spread out from there, they still keep their "citizenship." If reunited again after having been apart for a long time, they will still not fight but collaborate.

Exactly at what point a colony becomes a megacolony and a supercolony is a matter of definition and debate, but that such entities can span square kilometers and contain millions of queens and hundreds of millions of workers, is clear.

Huge Argentine ant colonies of this type are rare in their native range, but common abroad. One explanation for this may be that the fertile expatriates were initially relatively few in number, and therefore, the genetic base was narrow, meaning that after this squeeze at emigration, the immigrants ended up with a far smaller genetic variability than the home population. This would facilitate the formation of supercolonies but would also mean that there is less potential for genetic adaptation to changing environments and a larger risk for a population crash. In his book *Anthill*, the well-reputed formicologist, E. O. Wilson, pictures supercolonies as an evolutionary blind alley that is bound to be short-lived.

If so, even for the short run, supercolonies offer several significant competitive advantages over standard ant colonies. First, they spend far less time and energy on fights within the species and so can devote these resources to more productive ends. Second, they can amass a much higher number of ants in a given area and thereby better harvest available resources and fight-off competitors. Third, by being able to share resources, be it food or workers, between nests over a larger area, they can better cope with local disasters such as floods, fires, or shifts in the landscape caused by man's earth-moving machinery.

Ecological Effects of the Argentine Ant in the Mediterranean Area

Once the supercolonies had been identified, the speculation and reports of the devastation the enormous numbers of ants could cause started to appear, many of them, no doubt, finding inspiration from the horror stories of home invasions. Not only would the Argentine ants eradicate all other ant species, so the stories went, they would also impoverish the ecosystems of all other invertebrates and, moving up the food chain, subsequently of the vertebrates feeding on them. Plants, too, would suffer, especially those that native ants help spread the seeds of. Furthermore, like many other ants, the Argentine ones protect aphids and larvae of butterflies and moths with sugary deposits in order to collect and consume those, and when the massive armies of small ants do it, the protection from their natural enemies could get effective enough to allow those to mass-develop, thereby putting harvests at risk. Although these postulations were clearly alarmist, subsequent research from different parts of the world gave at least some credence to most of the allegations [19–23]. Partly because of this, ecological horror stories found sufficient support to survive in the popular literature. However, as James Wetterer demonstrated in his 2006 article [24], sloppy quotations and citations also helped to keep alarmist rumors alive.

Madeira, with its special position as being the first place outside of the Argentine ant's natural area of distribution, in which an invasion was reported, offers a rich example of claims of inflated negative impact. In the words of Wetterer [24] in his abstract:

> "Many authors have written about how exotic ants invaded the Atlantic islands of Madeira and negatively impacted or even completely exterminated its native ants, despite the lack of first-hand observations concerning such impact. I examine how quotation error (misrepresentation of previous work) and citation copying (citing unexamined publications referred to by others) led to the origin and spread of the erroneous story of ant extinctions in Madeira. Quotation error and citation copying may be more common than most scientists realize, particularly when authors cite references that are written in languages they do not understand."

No doubt he was provoked to write this when the study he and his coworkers made, based on firsthand observations [1], revealed a fairly limited elevation range of the invasion, and insufficient evidence of the elimination of native ant species, even from areas in which the Argentine ant was abundantly present.

However, there are also many other studies that result in a somewhat different picture. One from California, for example, concluded that the Argentine ant replaced most indigenous ant species, but that ant biomass remained more or less the same, and the effects on other ground-living arthropods of the invasive ants were very similar to effects the native ants had on them [25].

An undisputable effect, however, when *L. humile* invades a new area, is that the numbers of native ants are reduced, occasionally to the point of total local replacement. There are two main mechanisms for this process of reduction or elimination of other ant species. One is killing them in outright fights between individuals, or wars between colonies. The winning weapon of the Argentine ants in this lies in their numbers. Individually, they may be no match for their larger and often sting-armed opponents, but ten-or-more-to-one gives them an edge, enabling them to win.

The other and even more important way is through resource—primarily food—competition.

Observations, as well as experiments, show that Argentine ants are quick to find food and, again through their numbers, frighten and drive individuals of competing species away or kill them outright. As can be expected, they first and foremost disenfranchise the native species closest to themselves in habits and habitat preferences, with lesser impact of those most different. [26] Counted in the number of individuals, ants become more numerous, as the invader is smaller than most native species. Counting the total weight of the ants, however, their total biomass may not change that much.

The follow-on or cascading effects are real, but seldom as catastrophic as frequently claimed. They mostly relate to specific interactions of the replaced native ants with other species. If they are important prey for lizards or birds, for instance, these may not find the Argentine ant equally tasty. If the native ants drag plant seeds around and (accidently or by design) drop them where they might germinate, *L. humile* might not do the same. As an ant with a "sweet tooth" and with a higher density of workers, the Argentine ant is likely to do a better job of protecting aphids and larvae with sugary droppings than their native guardians would; this could help boost the suckers' numbers and, concomitantly, the damage they cause to vegetation, including agricultural crops. These effects, however, are seldom dramatic, at least not in Mediterranean environments.

Control of the Argentine Ant

As with other unwanted ants, a large number of physical and chemical agents have been tried to eradicate or control the Argentine ant, from digging up nests and encasing them in tar or cement to DDT treatment and cyanide fumigation. Many responsible authorities have come to the same conclusion as the Biosecurity Unit of the Ministry for Primary Industries in New Zealand, which says that the only effective method for control at present is hand-laying bait containing the slow-working toxin fibronil near-targeted ant nests. The toxic bait is carried back to the nest and fed to the queen, larvae, and other workers. The agency goes on to say that "The bait has no effect on humans, cats or dogs. Fibronil is the same ingredient used in flea control for cats and dogs. However, fibronil is toxic to fish and should not be used close to stormwater channels such as gutters, or close to streams or other waterbodies."

Obviously based on experience, it also adds, "Well-planned and executed baiting treatments can result in dramatic reductions in ant populations. But the tricky bit is always killing the last 1% of the population—that is where most of the budget and effort goes. Argentine ants are no exception to this rule. In most places, control is the most practical option, but in some locations, such as off-shore islands, complete eradication may be feasible."

On the other hand, the IUCN Invasive Species Specialist Group specifically warns against the application of the toxin in such places. Among general considerations with regard to *L. humile*, they say: "Non-target impacts must be weighed up carefully against the benefits of ant eradication. Cleary, treating whole ecosystems or islands is too risky as entire populations of rare invertebrates may be at risk of extinction."

There is currently no biological control agent that can be used against the Argentine ant, but scientists are working on various sophisticated ways to break up the super colonies and create infighting that might reduce the colonies to the small size and benign impact they have in their native range of distribution.

The most obvious such method could be to introduce the genetic variability from the home range to the supercolonies abroad. This has been proposed, among others, by Japanese researchers, as a way of controlling the Hokkaido supercolony. The disadvantage with increasing the genetic base of invasive populations is that their adaptive potential may increase instead of decrease, eviscerating the hope of a possible population collapse.

Research shows that ants use different types of "recognition chemicals" on their exoskeleton to identify friends and foes. The detailed structure of these linear hydrocarbons, with a few methyl groups as side chains, is inherited and is identical among related ants. Even within supercolonies with their millions of individuals, the recognition chemicals are similar enough to trigger positive identification. Alterations in these chemicals would transform "kissing cousins into mortal enemies," as Kenneth Shea, Robert Sulc, and Neil Tsutsui put it when they presented their idea and preliminary test results at the 232nd national meeting of the American Chemical Society in 2006. They synthesized different versions of the ants' recognition chemicals and sprayed them onto Argentine ants, thereby inciting fights to the death between the artificially coated ants and their nest-mates [27].

Other control ideas relate to the pheromone trail that ants secrete and follow as they navigate their way outside the nest. Synthesizing and spraying these compounds on the terrain all around a colony would cause the ants to lose their sense of orientation and with it the ability to synchronize their efforts. As pheromones are to a large extent species specific, application of the Argentine ant's pheromones would, so the theory goes, disrupt the colony's functioning and severely handicap them in the competition for resources with other ants.

Economic Aspects of the Argentine Ant Invasion

There seems to have been no comprehensive estimate made of the economic costs associated with the Argentine ant on Mediterranean shores. However, from the information provided by companies engaged in pest control, it seems that ants in homes and other buildings constitute the real bulk of ant eradication and control operations, and thereby probably also of the costs. Whether control efforts are directly linked to the economic damage of the ants or, in the case of private homes, more to the annoyance that their presence causes is difficult to assess. With regard to crop damage caused by helping insect sapsuckers to proliferate, there seems to be no data from the Mediterranean region.

In other parts of the world, notably New Zealand, some attention has been given to economic damage estimates. In a proactive assessment by Darren Ward, aimed at studying the potential damage the Argentine ant could do in the Hawke's Bay region of New Zealand if the ant were to invade the area, focus was put on the agricultural damage that increased ant-protected populations of sapsucking aphids and moth larvae could do to agriculture. The anticipated price tag for the relatively small region studied was put at NZ$23 million annually.

However, in a national assessment of the potential future economic costs, the same researcher and a colleague of his, Richard Toft, presented what appear to be significantly lower figures. After concluding that, there are generally two main economic impacts associated with Argentine ants: for spraying, nest destruction and other control efforts and for the loss of horticultural productivity. Full annual expenditure on control once Argentine ants have established throughout their predicted New Zealand range was estimated to be NZ$68 million per year in 2002 dollars, but given the slow rate of spread of the species, they estimated the 2002/2003 costs to be only NZ$0.6 million and projected the 2009/2010 to be NZ$1 million. Households, they wrote, account for the bulk of the cost (88%) [28].

They go on to say that they believe the main impact of Argentine ants to be their association with sapsucking insects but then admit that there is little information on Argentine ants affecting horticulture and no quantification of economic losses, either realized or potential. Similarly, although the potential exists for impact on the beekeeping industry, there have been no such reports in New Zealand, to their knowledge.

In Britain, the damage assessment is more toned-down. The Great Britain nonnative fact sheet says: "Economic costs are considered to be low but disruption to ecosystem services could contribute to economic losses. Argentine ants tend pest homoptera and protect them from natural enemies, therefore, causing increased crop damage."

The Mediterranean Supercolony: Current Population Trend

Soon after the discovery of the supercolony of Argentine ants on the northwestern shores of the Mediterranean, it was found that although the ants in the westernmost part, Atlantic Spain and Portugal, and the eastern-most part in Italy did belong to the same supercolony, there was a good stretch in between where another supercolony ruled. This one became known as the Catalonian supercolony. On the distribution map by Geraud et al. [18], the Catalonian supercolony is marked by open circles and the main supercolony with filled-in circles.

When the presence of the Argentine ant was reported from evermore places during the first part of the twentieth century, it was said that it "must be expected to penetrate the Eastern Mediterranean soon" [29]. More than 60 years later, that has not yet happened. *L. humile* has not yet been reported east of Crete [2]. This clearly indicates that the spread in the region has slowed down considerably and perhaps even come to a halt.

There are few, if any, studies of the current population trend for the whole area of distribution on the northwestern shores of the Mediterranean. Espadaler and Gómez updated the map with regard to the Iberian Peninsula in 2002 [30] but with regard to a time trend only concluded that there is no doubt that the ant has spread since its introduction.

Information on overall population densities within areas already colonized is also lacking. Generally, it can be said that the Argentine ant has been most successful in invading areas in which humans have disturbed the landscape and soil. On a local

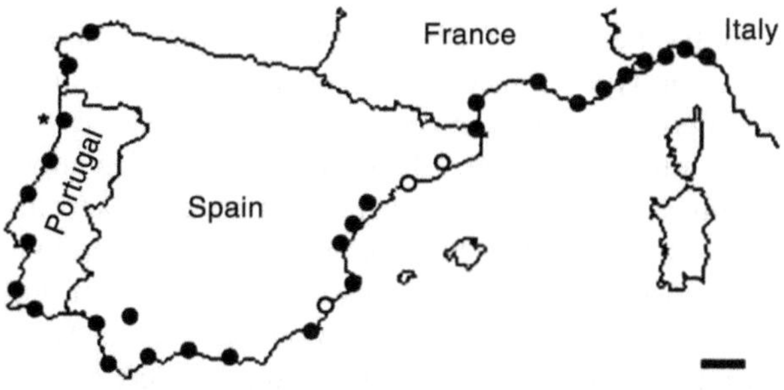

Distribution of the Argentine ant, *Linepithema humile*, in the Iberian Peninsula (Portugal and Spain) and the Balearic Islands, updated to 2002

level, new ant advances often follow new such developments. However, their advance has not been limited to such places; they have also moved into others, such as relatively intact forests. Also, generally speaking, they thrive best in marine near localities and seem not to have made major advances outside those areas. However, in some places, they have. To what extent such spread, in balance, is compensated by some thinning out of populations on "old" sites, as is occasionally reported, is not known.

In some reported situations, native ants have put up an effective resistance to the invading Argentine ant. The Italian island of Corsica is the most referred-to example of this. There, the dominant native ant, *Tapinoma nigerrimum,* seems to have managed to limit the ability of the Argentine ant to spread and invade. While *L. humile* would (mostly) win the setup interspecies fights, *T. nigerrimum* would by and large be more dominating, when it comes to food competition by being faster to locate ad hoc sources. It also often invaded nests of the Argentine ant, while the reverse has never been observed [31].

From other parts of the world, there are reports that may be relevant for the future of the Argentine ant in Europe in general and for the Mediterranean supercolony in particular.

In New Zealand, a population collapse of the Argentinean ant was reported on 2011 [32], two decades after the ant's introduction in 1990. In North America, the native winter ant (*Prenolepis imparis*) has been found to have a chemical defense that can keep *L. humile* in check and limit its expansion [33].

From USA, there is also a report of the Asian needle ant (*Pachycondyla chinensis*) gradually replacing the Argentine ant in an urban area. The competitive advantages of the Asian species over the Latin American one were thought to be its venomous sting and greater cold tolerance [34].

And another invasive ant species, the fire ant (*Solenopsis invicta*) has succeeded to replace the Argentine ant in the greater Austin area in Texas and has also weakened the population in several other parts of the state.

Summary

The Argentine ant has become a successful invader in many parts of the world that have a Mediterranean-type climate, but nowhere more than on the northwestern shore of the Mediterranean itself. A key to its invasive success is its tendency to form supercolonies outside its natural distribution area, which allows a much higher density of individuals than competing native ant species can muster. The largest known supercolony stretches 6000 km from northwestern Iberia via France to Italy and is thought to consist of several billion workers and tens of millions of queens. Over this area, it dominates the ant community and has replaced native species, even if not entirely driving them to extinction. Cascading effects on animals and plants, of the sort that has been documented elsewhere, are more or less self-evident, even as few have been reported from this area.

The initial explosive population development appears to have leveled off, but there are, as yet, no signs of a collapse. So, well over 100 years after its introduction, the Argentine ant is still ruling its Southern European fortress.

References

1. Wetterer, J. K., Espadaler, X., Wetterer, A. L., Aquin-Pombo, D., & Franquinto-Aguiar, A. M. (2006). Long-term impact of exotic ants on native ants of Madeira. *Ecological Entomology, 31*, 358–368.
2. Wetterer, J. K., Wild, A. L., Suarez, A. V., Roura-Pascual, N., & Espadaler, X. Worldwide spread of the argentine ant, *Linepithema humile* (Hymenoptera: Formicidae) Myrmecol. *Myrmecological News, 12*, 187–194.
3. Mayr, G. (1868). Formicidae Novae Americanae collectae a Prof. P. de Strobel. *Annuario della Societa dei Naturalisti Modena, 3*, 161–178.
4. Suarez, A. V., Holway, D. A., & Case, T. J. (2001). Patterns of spread in biological invasions dominated by long-distance jump dispersal: Insights from argentine ants. *Proceedings of the National Academy of Sciences of the United States of America, 98*(3), 1095–1100.
5. Stoll, O. (1898). Zur Kenntniss der geographischen Verbreitung der Ameisen. *Mittheil Schweize Entomol Gesellschaft, 10*, 120–126.
6. Way, M. J., Cammell, M. E., Paiva, M. R., & Collingwood, C. A. (1997). Distribution and dynamics of the argentine ant Linepithema (Iridomyrmex) humile (Mayr) in relation to vegetation, soil conditions, topography and native competitor ants in Portugal. *Insectes Sociaux, 44*, 415–433.
7. Schmitz, E. (1897). As formigas da Madeira. *Annales des Sciences Naturelles, 4*, 77.
8. Martins, M. N. (1907). Une fourmi terrible envahissant l'Europe (*Irodemyrmex humilis* MAYR). *Brotéria Series Zooligia, 6*, 101–102.
9. Silvestri, F. (1922). La Formica Argentina. *R. Laboratorio di Entomologia Agraria Portici, 1*, 1–7.
10. Chopard, L. (1921). La fourmi d'Argentine, *Iridomyrmex humilis* Var. Arrogans Santschi, dans le Midi de la France. *Annls Epiphyties, 7*, 237–266.
11. Marchal, P. (1917). La fourmi d'Argentine (*Iridomyrmex humilis* MAYR). *Bulletin de la Societé d'Etude et de Vulgarisation de Zoologie et d'Agriculture de Bordeaux, 16*, 23–26.
12. Frisque, K. (1935). La Fourmi d'Argentine *Iridomyrmex humilis* MAYR dans les serres en Belgique. *Annales de la Société Entomologique de Belgique, 75*, 148–153.
13. Chopard, L. (1921). La fourmi d'Argentine *Iridomyrmex humilis* Var. Arrogans Santschi dans le midi de la France. *Annls Epiphyties, 7*, 237–265.
14. Martinez, M. D., Ornosa, C., & Gamarra, P. (1997). *Linepithema humile* (Mayr, 1868) (hymenoptera: Formicidae) en las viviendas de Madrid. *Boletín de la Asociación española de Entomología, 21*, 275–276.
15. Donisthorpe, H. (1930). The ants (Formicidae) and guests (myrmecophiles) of Windsor Forest and district. *The Entomologist's Record and Journal of Variation, 42*, 13–16.
16. Carpenter, G. H. (1902). Household insects. A tropical black ant. *Iridomyrmex humilis*, MAYR. *Economic Proceedings of the Royal Dublin Society, 1*, 155–157.
17. Higashi, S., & Yamauchi, K. (1997). Influence of a Supercolonial ant Formica (Formica) yessensis Forel on the distribution of other ants in Ishikari coast. *Japanese Journal of Ecology, 29*, 257–264.
18. Giraud, T., Pedersen, J. S., & Kelle, L. (2002). Evolution of supercolonies: The argentine ants of southern Europe. *Proceedings of the National Academy of Sciences of the United States of America, 99*(9), 6075–6079.
19. Lach, L. (2003). Invasive ants: Unwanted partners in ant—Plant interactions? *Annals of the Missouri Botanical Garden, 90*, 91–108.

20. Rodriguez-Cabal, M., Stuble, K. L., Nuñez, M. A., & Sanders, N. J. (2009). Quantitative analysis of the effects of the exotic argentine ant on seed dispersal mutualisms. *Biology Letters, 5*, 499–502.
21. Savage, A. M., Rudgers, J. A., & Whitney, K. D. (2009). Elevated dominance of extrafloral nectary-bearing plants is associated with increased abundances of an invasive ant and reduced native ant richness. *Diversity and Distributions, 15*, 751–761.0.
22. Suarez, A. V., & Case, T. J. (2002). Bottom-up effects on persistence of a specialist predator: Ant invasions and horned lizards. *Ecological Applications, 12*, 291–298.
23. Vega, S. J., & Rust, M. K. (2001). The argentine ant—A significant invasive species in agricultural, urban and natural environments. *Sociobiology, 37*, 3–25.
24. Wetterer, J. K. (2006). Quotation error, citation copying and ant extinctions in Madeira. *Scientometrics, 67*(3), 351–372.
25. Holway, D. A. (1998). Effect of argentine ant invasions on ground-dwelling arthropods in northern California riparian woodlands. *Oecologia, 116*, 252–258.
26. Carpintero, S., Retana, J., Cérda, X., Reves-López, J., & Arias de Revna, L. (2007). Exploitative strategies of the invasive argentine ant (*Linepithema humile*) and native ant species in a southern Spanish pine forest. *Environmental Entomology, 36*(5), 1100–1111.
27. Retrieved from http://www.eurekalert.org/pub_releases/2006-09/acs-csf082806.php
28. Ward, D. F., & Toft, R. (2011). *Argentine ants in New Zealand.* Retrieved from http://argentine-ants.landcareresearch.co.nz
29. Bytinski-Salz, H. (1952). The zoogeography of the ants in the near east. *İstanbul Üniversitesi Fen Fakültesi mecmuası. Seri B, 18*, 67–74.
30. Espadaler, X., & Gómez, C. (2003). The argentine ant on the Iberian peninsula. *Sociobiology, 42*(1), 187–192.
31. Blight, O., Provost, E., Renucci, M., Tirard, A., & Orgeas, J. (2010). A native ant armed to limit the spread of the argentine ant. *Biological Invasions, 12*, 3785–3793.
32. Cooling, M., Hartley, S., Sim, D. A., & Lester, P. J. (2012). The widespread collapse of an invasive species: Argentine ants (*Linepithema humile*) in New Zealand. *Biology Letters, 8*(3), 430–433.
33. Sorrells, T. R., Kuritzky, L. Y., Kauhanen, P. G., Fitzgerald, K., Sturgis, S. J., et al. (2011). Chemical defense by the native winter ant (*Prenolepis imparis*) against the invasive argentine ant (*Linepithema humile*). *PLoS One, 6*(4), e18717.
34. Rice, E. S., & Silverman, J. (2013). Propagule pressure and climate contribute to the displacement of *Linepithema humile* by *Pachycondyla chinensis*. *PLoS One, 8*(2), e56281.

Dutch Elm Disease in Europe and North America

Around the turn of the nineteenth century into the twentieth, branches and entire elm trees began to die off on a large scale in Europe. Elms were then much appreciated. They often dominated parks and lined major boulevards and numerous other streets and were often planted as ornamental trees in gardens. The relatively sudden deaths were noted and much debated at the time, but the cause was unknown.

In the Netherlands, Johanna Westerdijk, associate professor in plant pathology at the Utrecht University at the time, was assigned the task of finding the cause of—and a cure for—the elm disease. One of her doctoral students, Bea Schwarz, identified the fungi *Ophiostoma ulmi* in the vascular canals of the diseased trees and claimed it to be the cause [1], which another coworker of Westerdijk's, Christine Buisman, later proved [2]. The Netherlands, being the country in which the causal agent was identified, gave rise to the moniker, "Dutch elm disease" (DED).

A Short Presentation of the Main Actors

Elm Trees: The Victims and Hosts

Elms are flowering trees of the *Ulmus* genus. They first appeared some 20 million years ago during the Miocene period and are spread over most of the northern hemisphere and in Southeast Asia, over the Equator and down to Indonesia. Dependent on definition, as there are many hybridizations and seed-sterile so-called microspecies with only vegetative propagation and local extension, there are some 30–40 species. Most of them are Asian, Europe has three elm species, and North America has eight. Elm trees are hermaphrodites. Their pollen is windborne, and their seeds are rounded and wind dispersed.

Elm trees can grow to be several 100 years old, and planted specimens in parks and at roadsides are often more than 30 m high and very stately; those found in

A. Jernelöv, *The Long-Term Fate of Invasive Species*,
DOI 10.1007/978-3-319-55396-2_12

forests are mostly much smaller. Earlier, elm wood was extensively used for wet applications, e.g., for boats, docks, and water-carrying pipes. The elm has also long been an appreciated ornamental tree.

This tree on the campus of the University of Guelph (Canada) has survived, despite its neighbors' succumbing to disease. (Photo from http://news.nationalpost.com/news/canada/cloned-ontario-tree-gives-hope-for-canadas-decimated-elms Courtesy of University of Guelph)

Elm tree leaves (Photo from Wikipedia)

Elm tree blossoms (Photo from Wikipedia)

Elm tree trunk (Photo from Wikipedia)

Elm trees normally flower when they have reached a height of at least 6 m. Mature trees will store growth hormones in the “twig crotches”—intersections of branches and trunks, growing at intervals along the trunk in preparation for the growth explosion of foliage and flowers. This may be an important factor in the epidemiology of DED.

The Beetles: The Vectors of the Spread

Like almost all trees, elm trees have their bark beetles from the family Scolytidae. In the context of Dutch elm disease, three species are important: the smaller European elm bark beetle, *Scolytus multistriatus*; the elm bark beetle, *Hylurgopinus rufipes* (eastern and central North America); and the banded elm bark beetle, *Scolytus schevyrewi* (western North America). Since more than a 100 years, the smaller European elm bark beetle has been present in the USA [3] and for 70 years in Canada (Ontario) [4].

The adult of the over-wintering generation of all three species of beetles, dark brown and about 2–4 mm long, emerge from under the bark of dead or dying elms from late April through June and move to feed on tender bark in twig crotches of nearby, healthy elm trees, mostly near their outside crowns. Some researchers believe that the plant growth hormones stored there are essential for the female beetle to start egg production [5]. This could explain why smaller and younger, sexually immature, elm trees are much less prone to beetle attacks.

The long sap vessels are open and functioning at this time of the year, and the feeding beetles will cut into these vessels and damage them. This is the way in

Female (**a**) and male (**b**) of the smaller European elm bark beetle (Photo from http://www.cabi.org/isc/datasheet/49212)

which the beetles themselves damage the trees, generally insignificantly so as compared to the impact of the DED fungi.

After feeding, the female adults move on to tunnel into the inner bark of recently dead, dying, or severely weakened elm trees. They lay their eggs in galleries parallel to the grain of the wood. The larvae, small, whitish, and grub-like, hatch and feed in the layer immediately beneath the bark of the tree. Their life cycle may be completed in 35–40 days. This first generation of adults produce a second generation of larvae, of which some enter a developmental diapause and overwinter, but a portion continues to develop and begins to emerge in August. These adults produce a third generation. Due to overlapping of generations, adults may be present almost continuously from April to October.

The Fungi: The Perpetrators

The causative agent of the Dutch elm disease is the fungus *Ophiostoma ulmi,* together with its close relative, *O. novo-ulmi*. These fungi live in the vascular system of the elm tree, and once inside, the fungal spores move upwards with the water or sap flow, while their threadlike mycelium grows in all directions, including downward to the roots. From there, it may spread to nearby elm trees through grafted root systems. Otherwise, the fungi depend on insect vectors to reach other trees. Elm bark beetles are by far the most important of such vectors. Whether as larvae or as wood-gnawing adults, they ingest and become carriers of the fungi, and as they move to other trees to eat or lay their eggs, they spread the fungi.

The elm trees react to the infestation by producing plug-forming structures called tyloses, but they cannot do so quickly enough to block the fungus. Instead, the ill-timed defense gums up the vascular system even more than the mycelium of the fungi and contributes to the die-off of branches and, ultimately, the entire tree. The whole process often takes less than a year.

Conidia (asexual stage) of *O. ulmi* (Photographs by L. Grand from http://www.ces.ncsu.edu/depts/pp/notes/Ornamental/odin18/od18.htm)

The relationship between *O. ulmi* and *O. novo-ulmi* has been the subject of much speculation since the latter, more aggressive, form appeared and largely took over.

Introduction and Spread

Ophiostoma ulmi

How and exactly when the Dutch elm disease came to Europe is not known and neither is it clear where it first arrived. In a discourse from 1658, it is noted that a disease with symptoms similar to those of DED was spreading through English elm hedgerows on Cyprus [6]. It can also be noted that the Latin name of the large elm bark beetle, *Scolytus destructor*, was given to it in 1795 after an elm tree die-off episode in Oxford, England. Except when carrying the *O. ulma* fungi, the beetle is not otherwise known to cause a major destruction of its host tree.

There is support for the claim that there were outbreaks of Dutch elm disease in 1819 and 1867 in England [7, 8] and in Belgium (Brussels 1836 and 1896, Ghent 1885–1886) [9]. Some sources claim it came to the Netherlands just before the turn of the century from the Dutch East Indies, today's Indonesia [10]. Others put the arrival just over a decade later in 1910 [11]. By and large, however, there is an agreement that the disease originated in the Himalayas, as elm trees there have a high degree of resistance to or tolerance of the fungi [12]. (The distinction between the two is not always made clear; resistance means that the fungus doesn't infect the tree so that the tree will not be a carrier of the disease, while tolerance implies that the tree gets infected but shows no symptoms or is at least not itself severely damaged but may act as a depository of the fungus.)

In 1927, the disease was identified in Hertfordshire in England, but in all likelihood it had already been present on the British Isles for several years by then. As on the European continent, the disease spread quickly and caused considerable loss of fully grown elm trees during the 1930s but gradually lost virulence so both the number of trees infected and the severity of the damage declined during the 1940s. In the 1950s, despite a number of severe local flair-ups, Dutch elm disease was largely viewed as a problem of no great importance [13].

Also on the European continent, from the end of the 1940s, Dutch elm disease was seen as merely a nuisance. Virus infections were seen as the cause for the demise of the strength of the *Ophiostoma ulmi* fungi, reducing its sporulation [14]. However, in Scandinavia, where DED reached Sweden only in the 1950s, *O. ulmi* remained infectious and spread, albeit slowly, in the south of Sweden [15]. It also reached the Oslo area in Norway in 1963 [16].

Concerning the arrival of Dutch elm disease to North America, there are also some divergent opinions. According to one source, *Ophiostoma ulmi* came with elm logs imported from France after WW1 [17], while others give 1928 as the year of introduction [18], or even the 1930s [19].

The first identification of *Ophiostoma ulmi* in North America was made in 1930, in samples from Cleveland analyzed by Christine Buisman at the end of her 1 year séjour at Harvard University [20]. The fungi supposedly came to Ohio with a shipment of elm wood from France, although some sources say the Netherlands [21].

From Ohio, Dutch elm disease spread to New Jersey, where trees were found to be dying in 1933. It reached New York in 1941 despite extensive efforts to keep it out. Detroit was hit in 1950 [22], Chicago by 1960, and Minneapolis in 1970 [23].

To the Pacific coast, where introduced elm trees were extensively planted along city and town streets, DED came later and reached Portland in 1993 [24].

The Dutch elm disease reached eastern Canada during World War II and spread to Ontario in 1967, Manitoba in 1975, and Saskatchewan in 1981.

Also in North America, the virulence of *O. ulmi* went down over time, and even if the westward spread continued, its speed was reduced and tree mortality diminished in many areas.

Ophiostoma novo-ulmi

In the 1960s, when the severe threat to elm trees from DED was seen as being over, even as it was still spreading westward in North America, the disease came back in a much more aggressive form. As *O. ulmi* was still active in some areas, mostly in North America, the presence of a new causal agent was not immediately noticed. It was first identified as "the aggressive strain" of *O. ulmi* in the late 1960s, when it had reached Britain [25], and only later recognized as a different species [26]. Many researchers hold that the second DED pandemic, caused by *O. novo-ulmi*, originated more or less at the same time in Ukraine-Moldova in Europe (subspecies *novo-ulmi*) and in the Michigan/Indiana/Ohio area in the USA (subspecies *americana*). This supposedly took place already in the 1940s, which means that some of the westward movement of DED after that in North America could have taken place with *Ophiostoma novo-ulmi americana* as the agent.

In 1967, a shipment of logs of rock elm, *Ulmus thomasii*, from Canada introduced the *O. novo-ulmi americana* subspecies to England and started the second DED pandemic there. Subsequently it spread into the Netherlands, France, Spain, and most other western and central European countries. At the same time, *O. novo-ulmi* moved steadily westward from its origin in Ukraine and reached the Netherlands around 1975. As a result, the two subspecies met and hybridized [27].

To Sweden, *O. novo-ulmi* arrived in 1980 and accelerated the spread of DED north up to the areas of Gothenburg and Stockholm. It arrived in Norway in 1981.

On the island of Gotland in the Baltic, the first case of the Dutch elm disease was identified in 2005 [28].

In Europe, Finland is the only country to which the Dutch elm disease has not arrived.

It also moved eastward into southwest and Central Asia, where separate introductions might have speeded up the process of spread. At the same time, subspecies *americana* spread across the North American continent, reaching the east and west coasts by the 1970s and 1980s, respectively [12].

While *O. ulmi* lost much of it virulence and had its sporulation reduced due to so-called d factors, naturally occurring virus-like agents, there is no sign (yet) of *O. novo-ulmi* being similarly debilitated [29].

With the risk of complicating the picture further, it should be mentioned that a third DED pathogen called *Ophiostoma himal-ulmi* was found in the western Himalayas in 1993 [30]. This species proved to be a very aggressive pathogen to European and American elm species but caused no wilt symptoms on *Ulmus wallichiana*, its native host.

Early Signs and Progress of the Disease

The first, most often noted sign of DED infection is commonly referred to as "the flag." It is a branch carrying only a few wilted leaves at the tip. Shortly before that, close observers could have seen a tell-tale discoloration, with leaves turning yellow, curling up, browning and quickly falling off (leaving only the flag), or turning dull green before dying and remaining attached to the twigs for several weeks. The shoot tips wither and drop off. Browning of the water-conducting vessels of the whole outer growth ring seen on the cross section and brown streaking in the direction of the grain on the twig after removal of the bark are further characteristic diagnostic symptoms for DED in affected twigs and branches.

When the infection is caused by vectors in early spring, the first signs may appear in just over a month, and the tree is often dead within just over a year. If the infection occurs during late summer and autumn, leaves will often only change color prematurely and have a normal bud set; however, the branch will be dead in the next spring and the tree often by the following autumn. Infections via root grafts usually kill a tree faster than infections caused by vectors, as the fungi are transmitted rapidly through the propagules and from there are distributed throughout the tree with the sap stream. Death can then occur within a matter of months [31].

Minor DED symptoms (yellowing leaves on a single elm tree branch) (Photo by R. Jay Stipes)

This row of elms was killed by DED, which moved through the root systems from one tree to the next (Photo by R. Jay Stipes)

Ecological Effects

Dutch Elm disease has had a major effect on the elm trees of Europe and North America. Of the previously ubiquitous old large elm trees, a precious few are left, as a large majority of them have succumbed to the fungus pathogen. According to the UK government, as many as 60 million elm trees died in Britain alone during the two episodes of infestations [32]. With a similarly high mortality on continental Europe, the total loss in Europe might well have been in the range of 200 million elm trees. The North American loss figure may be in the same ball park. According to the USDA, 77 million elm trees died in the USA between 1930 and 1980 [33]. (An article in the Washington Post from 2001 by Phil McCombs had the same number of dead trees, but gave the period as being 1930–1970 [34].) Adding the mortality in the last 36 (46) years plus the elm tree die-off in Canada, 200 million may be a possible figure for North America, too.

With such enormous losses, major ecological effects might be anticipated, but not much has been found and documented on the subject. A reason is, of course, that a large part of the lost elm trees were planted along streets, in parks, and gardens and constituted part of the urban environment. When the elms died, they were mostly replaced by "look-alike" trees of other species. The flora and fauna that did lose out were mostly the organisms closely related to elm trees, and these are neither that numerous nor that prominent. In a study of insect species found on various types of trees in Britain and European Russia, 82 and 81, respectively, were noted for elms, which are small numbers compared to those for, e.g., oak, willow, birch, hawthorn, poplar, and pine [35]. In another comparative study in the same article, of the number of species of *Lepidoptera* and *Coleoptera* associated with various trees in Sweden and Britain, those on elm trees are 46 and 69, respectively, again far below those registered for many other trees.

On continental Europe, a study named 79 species of insects said to be specialized on elm trees [36]. A large group of them were species that merely sought refuge in the deep cracks in the thick bark of the old elm trees, and in the absence of a rich supply of alternative coarse-barked old trees such as oak, these insects became less abundant when DED killed off the old elms. The few of them, like the small European bark beetle (*Scolytus multistriatus*), the lesser-spotted pinion moths (*Cosmia affinis*), the white-spotted pinion (*Cosmia diffinis*), and the white-letter hairstreak (*Satyrium w-album*), that as larvae feed exclusively on elm, have declined dramatically since DED struck.

In Britain, a large group of lichens have been identified as being most often associated with elm trees. One source says 161 [37] another "over 200" [38] species. The lichen community on elm bark is characterized by the colorful and species-rich *Xanthorion* family, of which the bright yellow, leafy *Xanthoria parietina* is the most well known. They often grow together with gray and brown species of the *Physcia* family. Even more specialized on elm trees are the sap-grove lichen (*Bacidia incompta*) and orange-fruited elm lichen (*Caloplaca luteoalba*) which have undergone a dramatic decline and are now not only red listed in much of Europe but are also classified as priority species under the UK Biodiversity Action Plan.

In forests, too, the real mass of older and larger elm trees has died. This, however, does not mean that the elm species have disappeared. The fecundity of elms, with their huge production of seeds and root shoots, has secured their survival, but as the trees become increasingly exposed to DED when they grow large and mature, what remains is mostly bushes and small trees. For fauna dependent on elm leaves, this can be good enough, while those that use the rough bark of older trees or the flowers have to find alternatives.

No extinction of flora or fauna has been associated with Dutch elm disease.

Economic Effects

As noted above, Dutch elm disease may be responsible for the deaths of up to 400 million elm trees over a period of roughly 100 years. Most of those killed have been large, mature, much-appreciated trees close to inhabited areas. In the USA, for insurance purposes, the value of an elm tree has been put at US$ 2500 [10, 39]. If that figure is also used to measure their economic value, the total loss would amount to US$ 1 trillion or an average of US$ 10 billion a year.

A special type of damage occurred when Dutch elm disease came to the Padana Valley in the Emilia region of Italy in 1930 and subsequently to Tuscany, Marches, Umbria, and Lazio. In Italy, the elm was used as a living support for the grapevines in vineyards. Their importance was such that land in the wine district of Lazio wasn't valued according to acreage but according to the number of elm trees on it [40]. When DED had killed off most of the elm trees, it was said that "the wine wedded to the elm has become a barren widow." The Italian wine industry suffered huge losses before the local, highly DED-susceptible elm was replaced by a DED-tolerant Siberian species. The lost production might have been in the range of 10–20 million hectoliters (hl), to the tune of US$ 100 million in today's currency.

In addition to the value of the dead trees themselves, the costs of protective measures are generally included in estimates of the economic damage. I have found no summary of those but have found a large number of time- and city-specific figures, mostly from the USA but also some from Britain, the Netherlands, and the rest of Europe. They are generally in the range of US$ 1–3 million per year in cities with 1–3 million inhabitants, which broadly correspond to a dollar per person and year. This generalization would thus give a figure of some US$ 900 million as the annual cost of DED protection in Europe and North America combined.

Control Attempts

Attempts to protect and preserve elm trees have involved one or more of the following measures: restrictions in transport of elm wood from contaminated areas to DED-free ones; removal and destruction of dead or dying branches and trees;

chemical control of the insect vectors of DED, the elm bark beetles; injection of fungicides into the contaminated trees or trees at risk; and production and propagation of DED-resistant or DED-tolerant elm plants.

The latter was in fact one of the very first approaches, one that Christine Buisman worked on in the early 1930s. Unfortunately, the strain she developed turned out to be very sensitive to another fungus, a *Nectria* species [41]. The work since then to develop elm plants resistant to or at least more tolerant of DED has followed two main tracks. One has been to look for the few surviving trees in areas of massive die-off and clone those on the assumption that not only chance made them survive but some properties they had. (The tree pictured above on the University of Guelph campus is such an example.) The other track builds on the fact that there are Himalayan elm species that are reported to be DED resistant and uses them to produce resistant hybrids. (Many elm species easily hybridize.) Over the decades, there have been many elm varieties produced by both methods, but most of them have been disappointments when, at maturation at an age of perhaps 20 years, the trees thought to have been resistant/tolerant succumbed to DED anyway. Another recurring problem has been that elm varieties presumably tolerant to Dutch elm disease have turned out to be highly susceptible to elm yellow, systemic infections of phloem caused by prokaryotic microorganisms called mollicutes [42, 43]. Many of the hybrids, especially those created between American and Asian elm species, don't achieve the tall, stately stature of the parent species but remain smaller and more bushy. Yet another problem is that the strains called resistant may in fact be only DED tolerant and act as local depositories of the fungi, from which susceptible native elm species can readily be infected when they reach maturity and form blossoms.

Today, much plant breeding is done through genetic modification, and the chances of success are probably much higher, but given the long time until maturity, the real test of their efficacy is still to come.

Restrictions on transport of elm wood, as well as of the vector-transmitting beetles, have failed to protect many areas despite considerable efforts in some places. Nevertheless, four main havens in which the elm trees, both native and introduced, are still protected are the Canadian West Coast, most of China, Japan, and Australia.

Removal and destruction of dead or dying branches are practiced in most contaminated areas where some elm trees are still present, and these efforts constitute a large chunk of the control costs. Where religiously carried out, it has been possible to substantially reduce the death rate among mature trees and thereby, for the time being, protect the stands of elm trees. Given the long time to maturity of the elms, the death rate from DED and other causes has to be at the annual level of a few percent or smaller to sustain the presence of mature elm trees. In many places it has been noted that elm hedges, even very old ones, have survived, when DED has wiped out practically all other elm trees. The reason seems to be that with constant cutting and trimming, the elms constituting the hedges seldom bloom and therefore remain unattractive to the female elm bark beetles. In some places this has inspired people to try to save even big trees by regularly cutting and trimming them—an arduous undertaking.

Spraying of an elm tree to kill the beetles (Photo from [26])

When the second wave of Dutch elm disease hit, DDT and other chlorinated insecticides were still seen as general saviors from all the ills of insects, and massive spraying campaigns were undertaken to kill the beetles. Since it is difficult to get at the larvae under the bark, and that imagoes developing during the whole of the spring and summer months meant that spraying had to be repeated over and over again, it became both labor intensive and costly—and in the end, not very effective.

A technical breakthrough in protecting and saving elms from Dutch elm disease in Canada came when the focus shifted from stopping the beetles to stopping the fungus. If the right fungicide and enough of it was injected into the tree's vascular system, then the spores introduced by a beetle wouldn't be able to germinate, so the tree could not become infected. Technically, this method has been relatively successfully and has saved tens of thousands of elms in Canada over the past three decades [44]. It has also been employed in other countries but not with quite the same enthusiasm as in Canada.

Also an already infected tree can be saved by injection of systematic fungicides, if the infection has not yet spread to the roots. The injected chemicals then move with the transpiration stream through infected sapwood that is still functional, stop the pathogenic action of the fungus, and allow the tree to ward off the infection. In addition, it is most important that the fungicide is completely distributed throughout the crown of the affected tree during the injection procedure. Of the more than 600 chemicals tested for this purpose [45], the two that have become the most widely used in the USA are thiabendazole hypophosphite and propiconazole. Reported success rates for these are 55–79% [46].

A drawback is the high cost of repeated, multi-branch injections.

In actual DED management today, a combination of removal and destruction of dead and dying branches and trees, individual protection of especially valuable trees through twig cutting and/or fungicide injection, together with planting of DED-resistant or DED-tolerant elm varieties, are often practiced.

Public Perception and Official Policies

There is absolutely no doubt that the public has a very negative opinion of Dutch elm disease. As deep as the appreciation, even love, of elm trees was and is, as witnessed by all the planted elm trees along boulevards and streets, and in parks and gardens in both Europe and North America, as well as the efforts undertaken to save these trees, so deep is the deprecation, even hatred, of the mortal disease.

Official policies practically everywhere and at all levels of government are to try to prevent the fungus and its vectors from reaching the areas where they are not present and to control them where they are. The acceptance of the very high annual control costs, decade after decade, bears witness to the seriousness of the ambition. There is, arguably, no other nonhuman pathogen on which so many resources are spent to combat it.

What Could Be in Store for the Future?

The first wave of DED subsided and disappeared after some 30 years, when accumulation of so-called d factors—a naturally occurring, virus-like agent—had reduced sporulation and virulence of *O. ulmi*. With regard to *O. novo-ulmi*, Brasier couldn't see any such tendency in 1996, a similar time after it was identified as having caused its first DED outbreak. The spread of d factors within a population is hindered by the production of ascospores, which are free of virus particles, and by vegetative incompatibility (vic) between fungal strains.

The genetics of the *Ophiostoma* species has been extensively studied, and the rapid evolution of in particular *O. novo-ulmi* has been well documented, including several hybridizations. These events and their effects on the vic system of *O. novo-ulmi* probably have a clear bearing on the persistent high fitness of the pathogen [47, 48]. The hybridizations occurring between the two subspecies of *O. novo-ulmi* in central and other parts of Europe, as noted in the section above on Introduction and Spread, contributes to the ongoing evolution of the Dutch elm disease fungi [49, 50]. It is thus not apparent that loss of virulence on part of this fungus will save the elm trees any time soon.

The better hope is that plant breeders will come up with a resistant elm with the desired characteristics, despite the setbacks hitherto, as noted above in the Control Attempts section. Until then, the costly and labor-intensive damage-reduction measures must be maintained, if the elm trees as we know them are to survive.

References

1. Schwerz, M. B. (1922). *Das Zweigsterben der Ulmen, Trauerweiden und Pfirsichbäume: eine vergleichende-pathologische Studie*. Open Access available from the Utrecht University Repository.
2. Retrieved December 4, 2016, from http://alchetron.com/Christine-Buisman-1319978-W
3. Chapman, J. W. (1910). The introduction of a European scolytid (the smaller elm bark beetle, *Scolytus multistriatus* marsh) into Massachusetts. *Psyche, 17*, 63–68.
4. Retrieved December 5, 2016, from http://www.exoticpests.gc.ca/es-details/insect/2988
5. Retrieved December 5, 2016, from http://www.treehealthcare.co.uk/dutch-elm-disease/
6. Browne, T. (1658). *The gardens of Cyprus*. London: Henry G. Bohn.
7. Rackham, O. (1986). *The history of the countryside* (p. 242). London: Phoenix Giant.
8. Jefferies, R. (1883). *Nature near London*. London: Chatto & Windus.
9. Meulemans, M., & Parmentier, C. (1983). Studies on *Ceratocystis ulmi* in Belgium. In *Research on Dutch elm disease in Europe* (p. 86). London: HMSO.
10. Hubbes, M. (1999). The American elm and Dutch elm disease. *Forestry Chronicle, 75*(2), 265.
11. Retrieved from https://en.wikipedia.org/wiki/Elm
12. Brasier, C. M. (1990). China and the origin of Dutch elm disease: An appraisal. *Plant Pathology, 39*, 5–16.
13. Peace, T. R. (1960) *The status and development of elm disease in Britain* (Forestry Commission Bulletin 33). HMSO, London
14. Brasier, C. M. (1996). New horizons in Dutch elm disease control. In *Report on forest research* (pp. 20–28). London: Forestry Commission, HMSO.
15. Retrieved December 10, 2016, from https://www.skogssallskapet.se/artiklar--reportage/artiklar/2008-09-16-ask-och-alm---borta-om-bara-tio-ar.html
16. Retrieved December 10, 2016, from https://snl.no/almesyke
17. Retrieved from http://www.dutchelmdisease.ca/history/
18. Retrieved from https://en.wikipedia.org/wiki/Dutch_elm_disease
19. Dutch Elm Disease and the American Elm. Pest alert, USDA, Forest Service, NorthEast Area NA-PR-05-99.
20. Retrieved December 1, 2016, from http://alchetron.com/Christine-Buisman-1319978-W
21. Retrieved December 5, 2016, from https://en.wikipedia.org/wiki/Dutch_elm_disease
22. Baulch, V. (2001, December 20). How Detroit lost its stately elms. *Detroit News*.
23. New York Times. (1989). Retrieved from nytimes.com/1989/12/05/science/new-varieties-of-elm-raise-hope-of-rebirth-for-davastated-tree.html?sec=health&spon=&pagewanted=all
24. Retrieved December 5, 2016, from http://koin.com/2016/12/01/portland-forces-homeowner-to-remove-trees-wont-pay/
25. Retrieved December 8, 2016, from http://www.cabi.org/isc/datasheet/37594
26. Brasier, C. M. (1991). *Ophiostoma novo-ulmi sp. nov.*, causative agent of current Dutch elm disease pandemics. *Mycopathologia, 115*(3), 151–161.

27. Jeng, R. S., Bernier, L., & Brasier, C. M. (1988). A comparative study of cultural and electrophoretic characteristics of the Eurasian and North American races of *Ophiostoma ulmi*. *Canadian Journal of Botany, 66*(7), 1325–1333.
28. Östbrant, I.-L. (2013). *Holländsk almsjuka Gotlands Län år 2005–2013*. Hemse: Skogsstyrelsen Gotlands distrikt.
29. Brasier, C. M. (1996). New horizons in Dutch elm disease control. In *Report on forest research* (pp. 20–28). London: Forestry Commission, HMSO.
30. Brasier, C. M., & Mehrotra, M. D. (1995). Ophiostoma himal-ulmi sp. nov., a new species of Dutch elm disease fungus endemic to the Himalayas. *Mycological Research, 99*, 205–215.
31. Stipes, R. J., & Campana, R. J. (Eds.). (1981). *Compendium of elm diseases* (p. 96). St. Paul, MN: American Phytopathological Society.
32. Retrieved December 6, 2016, from http://www.forestry.gov.uk/fr/HCOU-4U4JCL
33. Retrieved December 5, 2016, from https://agresearchmag.ars.usda.gov/1996/jul/elms
34. Retrieved from http://landscaping.about.com/cs/treesshrubs/a/american_elms.htm
35. Southwood, T. R. E. (1961). The number of species of insect associated with various trees. *Journal of Animal Ecology, 30*, 1–8.
36. Möller, G. (1993). Ulmenerhaltung aus der Sicht des Naturschutzes—Probleme und Möglichkeiten. In: J. Kleinschmit & H. Weisgerber (Eds.), *Ist die Ulme noch zu retten? Berichtsband des 1. Ulmensymposiums in Hann. Münden am 21.* und 22. Mai 1992.
37. Richens, R. H. (1983). *Elm*. London: Cambridge University Press.
38. Edwards, B. (2005). *Lichens and elm trees*. Back from the brink management series.
39. Retrieved December 6, 2016, from http://elmcare.com/disease/dutchelm/history_of_dutch_elm_disease.htm
40. Wilson, C. L. (1975). The long battle against the dutch elm disease. *Journal of Arboriculture, 1*, 107–112.
41. Heybroek, H. M. (1957). Elm breeding in the Netherlands. *Silvae Genetica, 6*(3–4), 112–117.
42. Cornell University, Collage of Agriculture and Life Sciences. (2001). Dutch elm disease: *Ophiostoma novo-ulmi*; *O. ulmi*. Information sheet.
43. Retrieved December 7, 2016, from http://aggie-horticulture.tamu.edu/archives/parsons/trees/elmyellow.html
44. Retrieved December 4, 2016, from http://www.dutchelmdisease.ca/history/
45. Stipes, R. J. (2000). The management of Dutch elm disease. In C. P. Dunn (Ed.), *The elms—Breeding, conservation and disease management* (pp. 157–172). Boston, MA: Kluwer Academic Publishers.
46. Stennes, M. A. (2000). Dutch elm disease chemotherapy with Arbotect 20-S and Alamo. In C. P. Dunn (Ed.), *The elms—Breeding, conservation and disease management* (pp. 173–188). Boston, MA: Kluwer Academic Publishers.
47. Brasier, C. M. (2000). *Intercontinental spread and continuing evolution of the Dutch elm research pathogens* (pp. 61–72). Boston, MA: Kluwer Academic Publishers.
48. Brasier, C. M. (2000). Viruses as biological control agents of the Dutch elm disease fungus *Ophiostoma novo-ulmi*. In C. P. Dunn (Ed.), *The elms—Breeding, conservation and disease management* (pp. 201–212). Boston, MA: Kluwer Academic Publishers.
49. Brasier, C. M. (2001). Rapid evolution of introduced plant pathogens via interspecific hybridization. *Bioscience, 51*(2), 123–133. 47 ref.
50. Konrad, H., Kirisits, T., Riegler, M., Halmschlager, E., & Stauffer, C. (2002). Genetic evidence for natural hybridization between the Dutch elm disease pathogens *Ophiostoma novo-ulmi ssp. novo-ulmi* and *O. novo-ulmi* ssp. *americana*. *Plant Pathology, 51*(1), 78–84. 39 ref.

American Mink in Europe

A model sporting a silver mink fur coat, Getty Images/Hulton Archive

In the early and middle twentieth century, "a fur coat—preferably mink—represented for many women in Britain, as well as in France, Germany and Italy, the height of luxury, the ultimate status symbol and object of desire. In popular culture, the ambition to own a fur coat became a defining quality of femininity, tantamount to a secondary sexual characteristic. Cartoons made play of this, as when Punch published 'The First Mink', a series of drawings by Alex Graham which showed women reluctant to take off their fur coats in summertime, at cocktail parties in warm rooms or even while doing the washing up. In one of these images a tiny girl with a big smile parades before a mirror in her mother's fur, its hemline trailing on the floor. If a woman were lucky, a rich husband or male protector would buy her a fur. If she came into money in her own right it surprised no one if the first thing she bought herself

A. Jernelöv, *The Long-Term Fate of Invasive Species*,
DOI 10.1007/978-3-319-55396-2_13

was a fur coat. Fur epitomized a dream of glamour, the height of fashion, a longing for comfort and success" [1].

To meet the demand for fur of the American mink, farming had started in the USA in the 1880s. Initially, mink farmers had to compete with the fur from trapped wild mink and cope with all of the problems of raising nondomesticated animals in cages, so profitability was generally low and farming failures common. As time went by, wild minks became less common, farmers learned more about how to treat the animals, and the animals became selected for those individuals that could best cope with cage life. All in all, mink farming became a profitable business, and when, some time after World War I, mink fur also became a sought-after luxury for European women, farms breeding American mink were established in Europe, too. As fish was the best and cheapest feed for the farmed mink, Europe's first American mink farms were established in Norway (1927), the UK (1929), and Iceland (1931). As with most farmed but not tamed animals, some mink escaped, and the first feral mink population became established in southwest Norway in 1930. The trend accelerated strongly during German occupation of the country in World War II, when many farmers released their animals, either unable to feed them or unwilling to let the occupiers take them. Something similar happened in Denmark.

In the Soviet Union, American mink were intentionally released in a large number of places starting in 1933, for the purpose of establishing populations for hunting. Many of these releases, including at least one in Lithuania, resulted in wild populations. The same thing happened in Belarus [2].

In the 1950s, escaped mink from farms formed feral populations in England, Scotland, and Wales, as well as on the Iberian Peninsula. Lately, reproduction of escaped mink in the wild has also been observed in Northern Italy [3].

Something About the Biology and Habits of the American Mink

Photo by Terry L. Spivey From Wikipedia

The American mink (*Neovison vison*)[1] belongs to the mustelid family of middle-sized predators. It is a semiaquatic species, spending some of its time in water and living to a large degree on a diet of fish, especially in winter, when the cold-blooded fish move slowly. The adults weigh from ½ kg to 1½ kg, and their bodies are 30–45 cm long, to which comes an equal length of tail that makes up 50% of their overall length. Males are generally larger and heavier than females. The mink has a dense winter fur, made waterproof by oily guard hairs and a thick underfur. Among wild mink, their coloration is generally dark brown. In captivity, a number of color mutations, including black, white, silver, and beige, occurred and have been maintained through selective breeding.

On land, mink run with a bounding gait; in water they swim using undulating trunk movements.

They have anal glands that are used for scent marking, either through defecation or by rubbing the anal region on the ground. The ill-smelling excretion from the glands can also be deployed for protection in a similar manner to that of the skunk.

Both males and females are territorial and live alone. There is hardly any territorial overlap between minks of the same sex, but extensive overlap between those of opposite sexes. Thus, a male might have four or even five females that claim part of his territory as part of theirs, and they in turn may have three or four males on different parts of their territory. (Males generally keep a larger territory than the females.) The preferred territory contains a broad littoral zone with rocks, fallen tree trunks, and dense plant cover, plus, of course, a stretch of water—be it in the form of a pond, lake, river, estuary, or canal.

Mink dens are almost always close to water. They consist of long burrows with twisty passages and multiple entries to the small, dry nesting chamber, which is only a foot or so in diameter, lined with fur, feathers, and straw [2, 4].

The American mink is promiscuous and does not form lasting pairs. During the short mating season of about 3 weeks, males often abandon their territories to look for as many females as possible with which to mate. This, of course, leads to fights with other similarly inclined males. The females are equally nonselective but stay put, waiting for the males to come to them. In the southern part of their distribution, the mating season is in February–March, and in the northern part mating normally takes place in April.

After mating, egg implantation is often delayed, resulting in a long gestation period (40–75 days), with actual embryonic development taking 30–32 days. The average litter size is four kits. The kits are born blind, weigh 5–6 g each, and have silver-white coats. It takes more than 3 weeks until they open their eyes and about 2 months before they are weaned off of their mother's milk, at which time they start hunting for themselves, at first mostly for insects. They remain with their mother until autumn, but well before winter they will start their independent life. They become sexually mature during their first spring, at about 10 months of age.

The American mink is an opportunistic carnivore, eating a large variety of rodents, fish, amphibians, crustaceans, and birds. Eggs of ground-nesting birds are

[1] An earlier name, *Mustela vison*, is used in many of the references.

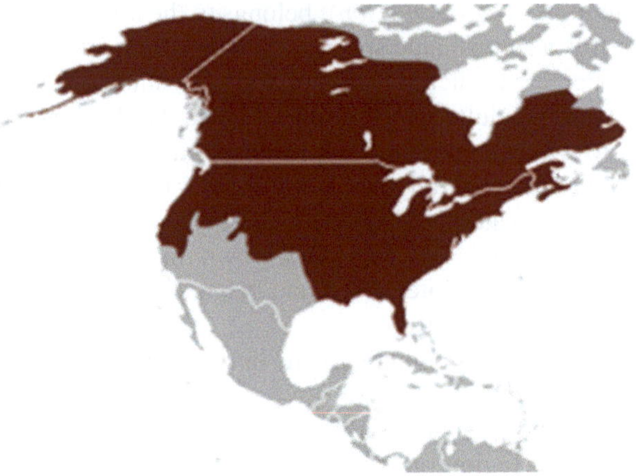

American Mink distribution in North America (Map from Wikipedia)

a favorite food when available, and insects and worms may do when nothing else is to be had.

The current native habitat of the American mink covers most of the North American continent, excluding the far south and the far north. Interestingly, its fossil ancestors have been found in practically the same range [5].

European Competitors to the American Mink

When the American mink established feral populations in Europe, it met competitors both with similar semiaquatic feeding habits and those with more pronounced specialities in their aquatic or terrestrial niches.

Closest to the American mink was the European mink (*Mustela lutreola*). Differences between the two species have a bearing on their intraspecies competiveness. The European mink is smaller than its American cousin, and its skull is less specialized for a carnivorous life with, for example, weaker dentition [6]. It does not prepare the same well-developed dens containing elaborated tunnel systems, and in water the European mink paddles with both front and back legs, making it a less-skillful swimmer [7].

A visual difference, probably without significance for competiveness, is the white markings on the upper lip that the European mink has, but the American lacks.

Already before the arrival of the American mink, the European one was in decline. A number of possible reasons for this have been proposed: wet land drainage, intensive hunting, and the loss of crayfish to the crayfish plague. When and where the expanding populations of feral American mink met the declining European mink population, they further pushed the natives out and contributed to their local extinction in some places.

The European polecat (*Mustela putorius*) is another European species that like the European mink is closely enough related to the American mink for them occasionally to mate, but the offspring, if the fetuses survive at all, are sterile.

The polecat mostly feeds on mice, voles and other small rodents, amphibians, and small ground-living birds. Fish is normally a rare item on its menu. Larger prey such as rabbits, hares, and geese are also regularly hunted.

Size-wise, the polecat almost matches the American mink, with males measuring 40 ± 5 cm and females 34 ± 5 cm and with weights of 1000–1500 g and 650–800 g, respectively. (Some significantly larger polecats have been recorded, but are thought to be hybrids with European mink.)

Compared to the American mink, the European polecat is less territorial, and its preferred territories are not necessarily waterfront ones.

European polecat (Photo from Wikipedia)

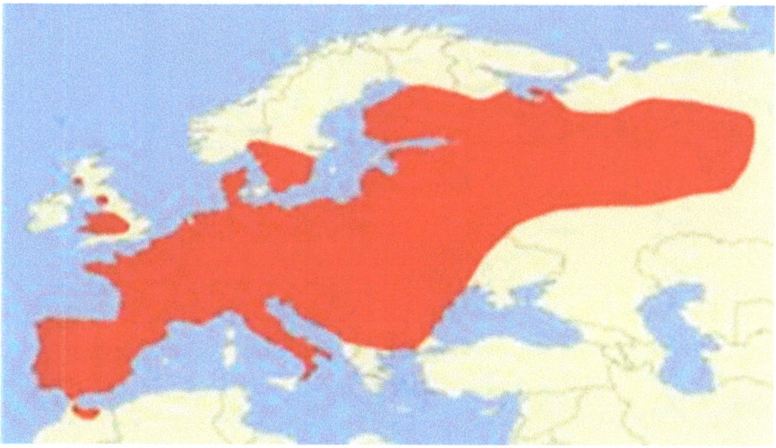

European polecat area of distribution (From Wikipedia)

Within its area of distribution, the polecat has for a long time been the subject of intensive hunting, both for its fur and to protect poultry. Nowadays, the polecat is itself protected in much of its range, and overall the population is not threatened, although locally it is declining.

The polecat can coexist with the European mink, but is reported to be outcompeted for food and driven out of wetland habitats by the American mink [8].

Arguably even more closely related to the American mink is the Siberian weasel (*Mustela sibirica*), but although the mink was introduced in some places such as around Novosibirsk and Irkutsk, along the Amur River and in Kamchatka [6], as the Siberian weasel's area of distribution is southern Siberia and China, the overlap between the species is outside the focus of this chapter.

Besides these close relatives, there are two other species that are the predominant competitors with the American mink in Europe: The Eurasian otter (*Lutra lutra*) and the red fox (*Vulpes vulpes*), both of which are significantly larger than the American mink.

The Eurasian otter can measure almost up to a meter in length and will typically weigh about 10 kg, although some males are much heavier than that. An excellent swimmer, it is also well adapted to aquatic environments, in which it mostly finds its diet of, predominantly, fish. However, crayfish, muskrats, small beavers, and even aquatic insects are also on its menu, together with frogs and water birds. Like the mink, they are strongly territorial and have fairly large territories, sometimes measuring several tens of kilometers, along the water's edge.

The otter has a wide distribution including Europe, Siberia, and China. In Europe, the densest populations are in Ireland, Latvia, the Shetland Islands, and along the Norwegian coast. They are rare or absent from the Alpine region.

In the 1960s the otter was hard hit by environmental pollutants such as PCBs in, for instance, Sweden, Central Europe, and Britain, after having been previously decimated through hunting. The populations have since been recovering and are still doing so in many places. When in competition with the otter, the American mink give up their aquatic hunting grounds and turn to terrestrial ones.

European red fox (*V. v. crucigera*), British Wildlife Centre, Surrey Photo from Wikipedia

The red fox is common throughout the Northern Hemisphere, from the Arctic Circle to North Africa, Central America, Northern India, and Southern China. It is a little shorter and lighter than otter, but still much larger than mink. It is a skillful hunter, with a solid and widespread reputation for being cunning. Small rodents are their favorite prey, but they are true omnivores. In the former Soviet Union, up to 300 animals and a few dozen plant species were found to be consumed by red foxes, including mink and weasels and prey as large as small deers [9]. The foxes often live in small family groups and are then territorial, using urine as a marker. When in competition with the fox, the mink give up the terrestrial hunting grounds and turn to aquatic ones.

The Further Spread in Europe

In Norway, in which Europe's first American mink farm had been established in 1927 and its first feral mink population was recorded in 1930, in the area of Sunnhordland, other escapees founded other wild populations as mink farming spread to other parts of the country. With the wartime addition of released minks, during which entire farms were emptied, an already large but scattered feral population received a further burst, and aided by the continued spread in the late 1940s and 1950s, the subpopulations met and continued to advance. By the 1980s, the minks were present all over the country except in the northernmost part called Finnmark. By the end of the millennium, however, they had also conquered this area and, in the very north along the coast, started to meet conspecifics from the old Soviet Union, in which 3700 minks had been released in 50 places in the Archangelsk and Murmansk regions in 1948 and had spread from there [10].

The minks in Norway are mostly brown, but other colors also appear in some wild populations as reminders of more recent releases of farmed animals. Besides occasions already referred to above, animal rights activists made further, significant contributions when, in the last decades, they stormed mink farms and "liberated" the animals.

In southwestern Norway, where the mink had first gained a foothold, the population began to decrease in the early 1980s. At the time, this was suspected to be a function of acid rain and the consequent decline of fish populations in acidified watercourses [11].

Since the turn of the millennia, the Norwegian mink population again appears to be declining, judging from catch statistics and road kills.

In some areas, the Norwegian mink population also spread into Sweden, although most of the Swedish population emanates from escapees from Swedish mink farms. The Swedish mink population saw exponential growth from the mid-1940s to the 1960s and then stabilized for two decades before exploding again and tripling to peak in the 1980s. After that followed a quick, decade-long drop and, thereafter, a steady, still ongoing decrease.

In a statistical study [12], appropriately published under the title, *When the fox is away, the mink will play*, the authors demonstrate how this development is coupled with the crash of the red fox population (due to an infection by the epizootic of sarcoptic mange, *Sarcoptes scabiei*) and its subsequent recovery. The study is based on data from different parts of Sweden over which the red fox population collapse occurred, with different lag times, as did the increase of the mink population. In a few districts, neither the infection of the fox population nor the explosion of the mink population actually occurred. To further test the connection, the authors added data on the population development of one of the fox's other prey, the mountain hare, *Lepus timidus*. Although not a mink prey, statistics were available for the hare, and it could conceivably represent fox prey species in general, many of which are also mink prey species. The conclusion was basically that when fox predation had all but vanished, the terrestrial prey species increased drastically in numbers, and the mink population tripled. When the foxes returned post-pandemic, prey species returned to their pre-episode numbers, and the mink the population followed suit. The slower, more continuous decline of the mink thereafter is attributed to the gradual recovery of the otters representing, from the point of view of the mink, the reemergence of an unbeatable competitor for limited aquatic food sources.

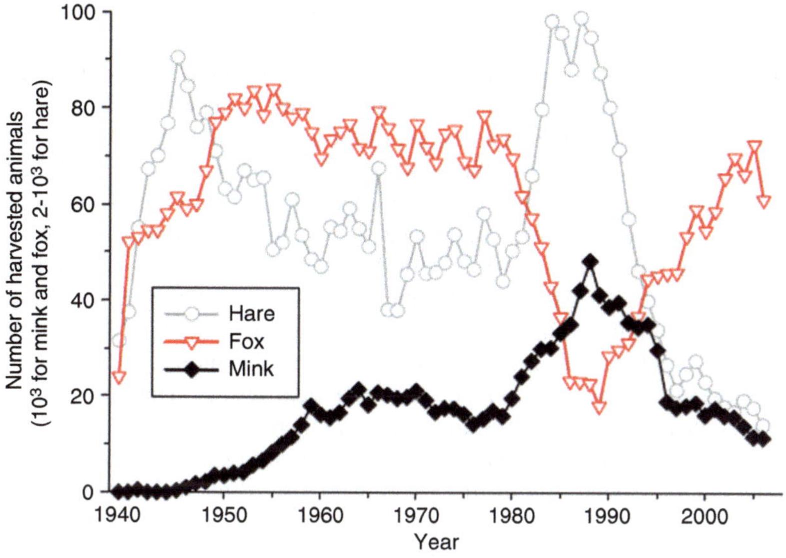

In Finland, the onset of the American mink invasion is somewhat unclear. The country had a number of American mink farms dating from the 1940s and onward, so escapees from those are prime suspects, but invaders from the Soviet Union/Russia and Sweden probably contributed over the years. The first observations of the American mink in the wild were in the 1930s in the southeast corner of the country, near the Soviet border, and they were in all likelihood immigrants from there. In the 1940s a feral population was established in the southwest, where mink farming was most intensive, and on islands in the archipelago there. This population expanded only slowly to the north and east. In the late 1950s and early 1960s, the spread accelerated, but little data is available for the next decades. What is clear is that in the early 1990s, the species was widespread in the country, with the highest densities in the east [13].

In Finland as in Sweden, the Eurasian otter population was much depressed in the mid-twentieth century and only slowly started to recover after the ban on chlorinated insecticides and PCBs in the 1970s. The presence of mink seems in no way have hindered the return of the otter, thereby contradicting the earlier claim that the mink could have contributed to the otter's decline. Conversely, the return of the otter has at least locally coincided with a reduction of the mink population.

Iceland was one of the pioneers of American mink farming in Europe, with its start in the early 1930s. Not long thereafter, the first American mink were observed in the wild. They spread over the island, and it is thought that the real bulk of the feral population originates from that first farm. Now they have expanded to cover all glacier-free parts of the island. The minks are being hunted, but less so nowadays, as the value of the fur from wild minks is low. Some international trapping enthusiasts come as tourists and pick up some of the slack, but some of them also use non-deadly methods and release their catches. Although registered catches are down, roadkills are few, and the mink is reputed as being hard to find, leading local zoologists generally to regard the mink population as stable.

In England, Scotland, and Wales, American mink farming also started early, but it was not until 1956 that the first confirmed breeding in the wild took place. By the end of 1967, the minks had spread to half of the counties in England and Wales and to the lowlands of Scotland. By the 1980s they had pretty much invaded the whole island. The population then may have exceeded 250,000 individuals, but soon after it started to decline over most of Britain due to interaction with the recovering Eurasian otter population [14]. In 2000, the American mink population was estimated to number 110,000 [15, 16] and in 2004 to number 36,950 [17]. Such a reduction in the number of mink as these figures suggest would indeed be dramatic, but counting mink is not easy, and differences in techniques or randomness in uncertainties could have exacerbated the numerical results. The downward trend, however, is clear and has also been revealed in other studies.

Using data from the National Otter Survey covering the period from 1977 to 2002, the authors found that of the 32 50 × 50 km^2, each with a minimum of 20 independent sites, the mink decreased in 60% and increased in 19% of the squares, with the opposite development for otter. Generally speaking, the mink declined in the north and west, held its position in the east, and increased in the Thames River valley in the center. The study also concluded that the return of the otter was the prime cause of the retreat of the mink [18].

In another study, trends in the distribution of native otter and invasive mink in northeast England were examined using a novel regression modeling approach, to analyze presence/absence data from field surveys, corroborated by contemporary predator-culling records. Between 1991 and 2002, the percentage of sites where mink signs were found decreased from 80% to 20%, while otter signs increased from 18% to 80%.

Annual indices of mink captures on shooting estates increased between 1980 and 1996, but were followed by a decline thereafter. Indices of the incidence of native otters were significantly related to those indicating the decline or displacement of invasive mink [19].

During the 1990s, animal rights activists performed a number of spectacular actions which might have "liberated" as many as 10,000 American minks from mink farms. Even that number, though, did nothing to reverse the overall decline of the feral mink population.

On Ireland, as in England, the first feral American mink populations were noted in the 1950s. The details of the spread are not known, but in the mid-1980s, it was classified as "widespread" [20], although described again later with the addition, "but not particularly abundant" [21]. Ireland has one of Europe's densest populations of otters, and that may be a reason for the mink's relative lack of success.

In Spain, France, and Germany, the American mink is present in several unconnected feral populations, some of them strong and expansive, such as the ones in the mink-farming districts of Brittany (France) and Galicia (Spain); others are weak and merely hanging on.

In Spain, the invasiveness of the feral American mink is seen as directly coupled to the presence or absence of competitors such as otters, European mink, and polecats. At least three of the, at that time, six feral mink populations in Spain had their origin in massive escapes through accidents or deliberate "liberations" from mink farms [22]. Later, the populations in Central Spain expanded and have since united into a single large population.

In France, besides in Brittany, there is also a significant population of American mink in the southwest along the Atlantic coast and near the Spanish border. This population is seen as a special threat to the remaining, fragmented, population of European mink [23].

In Germany, the American mink is mostly present in the north and east. As mink farming has not been much practiced in the country, migrating minks from Denmark and Poland are thought to have been the origin of the German populations.

In Poland, Lithuania, Latvia, and Estonia, as well as in the Russian enclave of Kaliningrad, the American mink is well established, especially along the Baltic Sea with its coastal lagoons and along rivers. These populations are linked to mink populations in Russia and Belarus.

From the Baltic states, there are several reports from hunters and naturalists of a recent decline in the number of mink, but there are yet no scientific reports to verify this [24].

Denmark is in many ways a special case. It has the world's largest production of farmed American mink, with well over a third of global production. Escapes and releases have occurred at a large scale and are occurring, and feral populations of mink are large and widely distributed over the Danish isles. However, a very large proportion of the existing feral population, almost 80% in one study, consists of animals that were born in captivity [25].

Ecological and Economic Effects

What Mink Eats

A large number of studies have been done on the feeding habits of American mink in Europe. The general conclusion is that feeding habits vary over the year and are strongly dependent on the degree of competition in different geographic areas. In Norway, in the presence of otter, the summer diet was dominated by crayfish and frogs and also included birds and some smaller mammalians such as mice, voles, and muskrat. In winter, the diet was primarily composed of small mammalians and birds. Early in spring, bird eggs and hatchlings top the list [26].

In a study of food habits of the American mink based on scat samples collected at three otter-free lakes in northeast Poland [27], it was found that although the mink feed on a large variety of prey, amphibians and fish dominated the diet during all seasons. The most common amphibian was the common frog, *Rana temporaria*, at one lake making up over 80% of the prey biomass. The fish consumed was mostly cyprinids and percids. In two of the lakes, fish were hunted mostly in winter, and in a third lake, they were hunted in summer, giving, for the whole area, a fairly stable average proportion of fish in the mink diet. Crayfish was an important food item in spring and summer. Birds, mammalians, and insects were supplementary food items for most of the minks, but for some individuals, with dens near bird colonies, eggs and birds dominated in May and June.

A study was conducted during summer of the food habits of mink living on islands in the Finnish archipelago. Otter were absent, but foxes were present on the larger island. The results showed that fish were the most important food item, comprising 33%, followed by birds, 28%, and small mammalians, mostly voles, at 15% [28].

A British study of mink feeding habits in three areas in southwest England, with foxes present but no otters, revealed that fish made up 42% of the annual diet, with more in the winter than in the rest of the year. Birds, especially moorhens (*Gallinula chloropus*), constituted 21%, with a much higher proportion in the spring and early summer, and mammals, 25%. Amphibians, crustaceans, and insects made up the rest of the diet [29].

In Spain, a 4-year, year-round study was conducted on feeding habits of mink, otter, and the spotted genet, *Genetta genetta*. It was found that the diets of the mink and the genet were dominated by the American crayfish, *Procambarus clarkii*, while the otter fed almost exclusively on fish. Compared to the mink, the genet seemed to specialize more on small mammals, while the mink preyed more on over-wintering aquatic birds, especially in autumn and winter [30].

A Personal Experience

In the 1970s and 1980s, I was myself engaged in aquaculture and, among other aspects of it, fish farming in large, earthen, so-called carp ponds in a forested area in southern Sweden. At the same time, my parents had a summer house in the archipelago of Stockholm, the surroundings of which I knew very well having spent a lot of time there in my childhood and youth. In both places American mink started to appear in the 1970s and by the 1980s had become a major problem.

Around the carp ponds, mink traps were regularly set. The annual mink catch in the 100 ha area around the ponds was a dozen or so in the 1970s, but exploded to well over a hundred per year in the 1980s. Needless to say, the minks took a heavy toll of the farmed fish and practically broke the economic spine of the enterprise before the population practically collapsed, when both foxes and otters returned. At its peak, the mink also replaced the fox as top predator on free-ranging chickens in the area.

In the archipelago, in an area previously designated as restricted for military use and with few summer houses, mink annually ransacked a few eider nests on the islands in the 1970s. By the 1980s, they were sufficiently numerous

and aggressive to scatter tern colonies, forcing the birds to abandon the practice of nesting in dense colonies on small islets and to move to individual nesting high above the water line on larger islands. Here, the decline of the mink population after the 1980s was much more gradual, and it took until well into the 2000s before colony nesting was again the norm for the terns and several other species of waterfowl.

Effects of Mink on Other Species and Vice Versa

The two effects of a large mink population, as illustrated by this personal experience from Sweden, are recurring themes in a number of studies from other countries. Damage to aquaculture features, in reports from all the Nordic countries and Scotland, and ground-nesting birds are the single-most mentioned group suffering from mink predation, in an overview by Bonesi and Palazon [15]. Reports from the Czech Republic, Estonia, Finland, Iceland, Ireland, Latvia, Lithuania, Poland, Sweden, and the UK all provide examples of damage to birds. Adaptive changes in nesting behavior among seabirds challenged by mink also figure in reports from Iceland and Scotland.

In the UK, a 97% decline in the water vole, *Arvicola amphibius*, since 1900 has been associated with the mink, for which the vole is a preferred prey [14]. On a local scale, this has been directly coupled to the presence of mink [31]. Studies in Belarus also show that the American mink negatively affects the water voles [32], while on islands in the archipelago in Finland, the populations of the bank vole, *Clethrionomys glareolus*, and field vole, *Microtus agrestis*, declined significantly when the mink arrived and recovered when it was removed in a local eradication campain [33].

Minks' preference for and dependence on crayfish has been noted in several countries, in different ways. In Ireland, mink abundance is closely-linked to that of the white-clawed crayfish *Austropotamobius pallipes* [21]. In the Czech Republic, the rebound of the stone crayfish, *Austropotamobius torrentium*, was noted after the experimental local removal of mink [34].

In northwest Portugal, American mink immigrants from Spain, after a slow spread south over two decades, suddenly accelerated their pace of penetration almost tenfold; the explanation found was that the new area they had entered was very rich in red swamp crayfish (*Procambarus clarkii*) [35].

In Spain, a study found that the American mink was "tropically subsidized" by an invasive crayfish occurring there in high densities. The mink reduced the size of its territories and became both more highly concentrated and more resilient in the face of control measures, since they had access to a rich food source that other predators hardly utilized [36].

Interactions between the European mink and the American mink, to the detriment of the former, are often discussed. It is feared that the invasion of the American mink may lead to the extinction of its European cousin, at least regionally. Often, the opposing population trends—the spread of the invader and retreat of the native—are referred to as support for the fears of competition extinction. Although it may appear to be a reasonable assumption to blame the negative trend of the European mink on competition with the newcomer, there are few actual studies to support it. In fact, a closer look at the population trends for the two mink species in different localities generally reveals that the decline of the European mink occurred prior to the arrival of the American mink, for example, in Finland, or that "anthropogenic factors, mainly changes in habitat quality and trapping, are the main cause of the decline of the European mink," as was the case in France [37].

In an experiment in Spain, however, when European mink were released into an area in which American mink were already present, the European mink had a low survival rate, with some of them in all likelihood being killed outright by male American mink [38]. Observations and an experiment in Belarus also show the decline of *Mustela lutreola* as a function of the competition from the invading *Neovison vison* [32, 39].

In Finnish studies, a negative effect has also been documented on the population of an amphibian, the common frog, *Rana temporaria*, due to mink predation [40]. There are, however, a number of studies that did not find any significant (beyond normal population variability) large-scale effect on prey species from mink predation from, for example, Denmark [41], Germany [42], and Ireland [21].

Other species that have been noted to have an effect on mink populations are the white-tailed eagle (*Haliaeetus albicilla*) [43, 44] and the beaver (*Castor fiber*) [45]. The eagle preys on mink and catches individuals especially when they are swimming, thereby forcing the mink to adapt and reduce mobility. The big birds also compete with minks for some prey species. However, the eagles are rare, even if many northern populations are recovering from near extinction some decades ago, and thus, they have little overall impact on the minks. Beaver dams change the landscape in ways benefitting the minks and provide additional opportunities for den constructions. There is a positive correlation between beaver and mink densities, despite that minks have been observed to prey on young beavers.

Economic Effects of Mink

Some studies have been done on the economic effects of mink. They all seem to conclude that although significant damage can be done to fish farms and poultry, the effect is local and not significant on a national level [46].

Divergent Trends Between Farmed and Feral Minks

When farming of American mink started in Europe, the minks were generally brown, and adult males would weigh 0.8–1.6 kg and females 0.5–1 kg. This was just some 20% over the weight of wild minks in America.

Today, few minks with the original brown fur are farmed in Europe. Over 100 years of domestication, together with higher prices for color mutants, spurred selective breeding to develop a wide variety of pelt colors, many of them either rare or unknown in nature. These include white and a host of shades of brown and gray, sometimes with tinges of blue or pink and bearing such exotic names as "lavender hope," "sapphire," "gun metal," and "mahogany." But the most popular pelts of all are "black." The popularity and thereby production varies strongly over time, but the black pelts mostly account for half of the overall market [47].

Selective breeding and domestication has also led to significantly larger animals. The average weight varies with color and sex, but today the average weight for all colors of female farmed mink is 1.4 kg, with a range of 1–2 kg, while the larger males average 2.7 kg with a range of 2–4 kg [48].

Feral mink populations often receive a continuing influx of new escapees from farms. In some cases, as in Denmark, that influx can be overwhelming. In other places without ongoing mink farming, it can be almost zero. This will have an effect on the genetic makeup of the local populations, most easily seen in the color and size of the animals. Natural selection, however, strongly favors the original brown color, so color mutations normally do not last long unless they are minor. Natural selection in European environments also seems to favor a smaller size, as can be seen from the fact that feral mink populations without a continued influx of farmed animals slowly but progressively produce smaller animals, eventually approximating the size of wild minks in America.

Control and Extermination Attempts

A number of methods have been tried to control and even exterminate feral mink populations locally and regionally. Mostly, traditional hunting methods such as trapping and shooting have been practiced, but attempts have also been made with poisoned bait, habitat manipulation, promotion of native competitors, and diseases like tick-spread tularemia. By far the best results have been obtained with intensive trapping, but also with this method, extermination has only succeeded on some small, offshore islands.

The Future

It seems likely that the slow spread of the American mink into southeastern Europe will continue; that the isolated populations in Spain, France, and Germany will expand their territories and ultimately unite; and that the established populations in Northern Europe will thin out further in the face of increasing otter populations.

The situation in the past in which there were empty niches due to the virtual absence of otters, low densities of foxes, and few other competitors, that allowed the American mink population to explode, is not likely to be repeated.

Summary

American mink that escaped or were released from fur farms have established feral populations in most European countries. Across northern Europe, where the process started in the late 1920s and 1930s, the mink is widespread on Iceland, Ireland, and the UK, as well as in Denmark and in an area stretching from Norway over Sweden, Finland, northwest Russia, and the Baltic states, to Poland and Belarus. This northern population is still expanding its territory: into Germany from the east and Ukraine from the west, as well as south into the Czech Republic, Slovakia, and Austria. On the Iberian Peninsula, in France and Italy, the American mink is more of a newcomer, as feral populations were established from the 1950s to the 1980s. In these countries the American mink does not have a countrywide distribution but exists in several isolated local populations, some of which are expanding their range and others that are just hanging on.

The American mink is semiaquatic and finds its prey both in aquatic and terrestrial environments. When it is the dominating predator in at least one of them, the population can expand quickly, but faced with superior competitors in both, such as the otter in water and the fox on land, the mink might just barely be able to survive. At the time when the American mink first came to northern Europe, the otter population had been at a historic low. This was first because of intensive hunting and afterward due to exposure to PCB and chlorinated insecticides. With the otters slowly but steadily coming back, American mink populations are declining in the UK, on the Scandinavian Peninsula, and in Finland and the Baltic states.

Where they exist in large numbers, American mink can devastate sea bird colonies and have a strong negative impact on populations of vole and wetland birds. They are also seen as a serious threat to the remaining populations of the European mink.

Locally, they can have a negative economic impact on poultry farming and aquaculture. Intensive trapping can significantly reduce mink numbers locally, but extermination campaigns have only been really successful on some small, offshore islands.

The current trends, with a slow spread of the mink into new areas in southeastern Europe; a slow expansion of the isolated populations in Spain, Italy, France, and Germany that will ultimately meet and unite; and a thinning out of the established populations on the British Isles and in Northern Europe as the otter continues its comeback, are likely to continue for the foreseeable future.

References

1. Dyhouse, C. (2011). Skin deep: The fall of fur. *History Today, 61*(11).
2. Harris, S., & Yalden, D. (2008). *Mammals of the British isles* (4th Revised ed.). Southampton: Mammal Society.
3. Spagnesi, M., Toso, S., & De Marinins, A. M. (2002). Mammafen d'Italia Ministero dell'Ambiente e della Tutela del Territorio e Istituto Nazionale per la Fauna Selvatica, Italy.
4. Seton, E. T. (1909). *Life-histories of northern animals: An account of the mammals of Manitoba*. New York: Scribner.
5. Kurtén, B. (1980). *Pleistocene mammals in North America*. Columbia University Press: New York.
6. Heptner, V. G., & Sludskii, A. A. (2002). *Mammals of the soviet union. Vol II, part 1b, carnivores (Mustelidae and Procyonnidae)*. Washington, DC: Smithsonian Institution Libraries and National Science Foundation.
7. Youngman, P. M. (1990). *Mustela lutreola. Mammalian Species, 362*, 1–3.
8. Sidorovich, V. E., MacDonald, D. W., Kruuk, H., & Krasko, A. (2000). Behavioural interactions between the naturalized American mink *Mustela vison* and the native riparian mustelids, NE Belarus, with implications for population changes. *Small Carnivore Conservation, 22*, 1–5.
9. Heptner, V. G., & Naumov, N. P. (1998). *Mammals of the soviet union: Vol II part 1a Sirena and Carnivora (sea cows, wolves and bears)*. Washington, DC: Science Publishers.
10. Lever, C. (1985). *Naturalized animals of the world*. Essex: Longman Group Ltd.
11. Bevanger, K., & Aelbu, O. (1986). Decrease in a Norwegian feral mink *Neovison vison* population—A response to acid precipitation? *Biological Conservation, 38*, 75–78.
12. Carlsson, N. O. L., Jeschke, J. M., Holmqvist, N., & Kindberg, J. (2010). Long-term data on invaders: When the fox is away, the mink will play. *Biological Invasions, 12*, 633–641.
13. Kauhala, K. (1996). Distributional history of the American mink (*Mustela vison*) in Finland with special reference to the trends of otter (*Lutra lutra*) populations. *Annales Zoologici Fennici, 33*, 283–291.
14. Jefferies, D. J. (2003). *The water vole and mink survey of Britain 1996–1998 with a history of the long-term changes in the status of both species and their causes* (p. 234). Ledbury: The Vincent Wildlife Trust.
15. Retrieved from http://www.theguardian.com/environment/2008/jun/07/wildlife.mink
16. Retrieved from http://www.gwct.org.uk/wildlife/research/mammals/american-mink/
17. Retrieved from http://www.fishforth.co.uk/inns/species/american-mink/
18. Bonesi, L., Strachan, R., & Macdonald, D. W. Why are there fewer signs of mink in England? Considering multiple hypotheses. *Biological Conservation, 130*(2), 268–277.
19. McDonald, R. A., O'hara, K., & Morrish, D. J. (2007). Decline of invasive alien mink (*Mustela vison*) is concurrent with recovery of native otters (*Lutra lutra*). *Diversity and Distributions, 13*, 92–98.
20. Smal, C. M. (1988). The American mink *Mustela vison* in Ireland. *Mammal Review, 18*, 201–208.
21. Smal, C. M. (1991). Population studies on feral American mink *Mustela vison* in Ireland. *Journal of Zoology (London), 224*, 233–249.
22. Ruiz-Olmo, J., Palazon, S., Bueno, F., Bravo, C., Munilla, I., & Romero, R. (1997). Distribution, status and colonization of the American mink *Mustela vison* in Spain. *Journal of Wildlife Research, 2*, 30–36.
23. Léger, F., & Ruette, S. (2005). Le vison d'Ame'rique, un espece qui se developpe in France. *Faune Sauvage, 266*, 29–36.
24. Bonesi, L., & Palazon, S. (2007). The American mink in Europe: Status, impact and control. *Biological Conservation, 134*(2), 470–483.

25. Hammershøj, M., Pertoldi, C., Asferg, T., Møller, T. B., & Kristensen, N. B. (2005). Danish free-ranging mink populations consist mainly of farm animals: Evidence from microsatellite and stable isotope analyses. *Journal for Nature Conservation, 13*, 267–274.
26. Retrieved from http://no.wikipedia.org/wiki/Mink
27. Brzezinski, M. (2008). Food habits of the American mink, *Mustela vison*, in the Mazurian Lakeland, NE Poland. *Mammalian Biology, 73*(3), 177–188.
28. Salo, P., Toivola, M., Nordström, M., & Korpimäki, E. (2010). Effects of home-range characteristics on the diet composition of female American mink in the Baltic Sea archipelago. *Annales Zoologici Fennici, 47*, 111–122.
29. Chanin, P. R. F., & Linn, I. (1980). The diet of the feral mink (*Mustela vison*) in southwest Britain. *Journal of Zoology, 192*(2), 205–223.
30. Melero, Y., Palazón, S., Bonesi, L., & Gosàlbez, J. (2008). Feeding habits of three sympatric mammals in NE Spain: The American mink, the spotted genet, and the Eurasian otter. *Acta Theriologica, 53*(3), 263–273.
31. Aars, J., Lambin, X., Denny, R., & Griffin, A. (2001). Water vole in the Scottish uplands: Distribution patterns of disturbed and pristine populations ahead and behind the American mink invasion front. *Animal Conservation, 4*, 187–194.
32. Macdonald, D. W., Sidorovich, V. E., Anisomova, E. I., Sidorovich, N. V., & Johnson, P. J. (2002). The impact of American mink *Mustela vison* and European mink *Mustela lutreola* on water voles *Arvicola terrestris* in Belarus. *Ecography, 25*, 295–302.
33. Banks, P. B., Norrdahl, K., Nordström, M., & Korpimäki, E. (2004). Dynamic impacts of feral mink predation on vole metapopulations in the outer archipelago of the Baltic Sea. *Oikos, 105*, 79–88.
34. Sálek, M., Poledník, L., Beran, V., & Sedlácek, F. (2005). The home ranges, movements and activity pattern of the American mink in the Czech Republic. In *IX International Mammalogical Congress, Sapporo, Japan*.
35. Rodrigues, D. C., Simoes, L., Mullins, J., Lampa, S., Mendes, R. C., Femandes, C., Rebelo, R., & Santos-Reis, M. (2014). Tracking the expansion of the American mink (*Neovison vison*) range in NW Portugal. *Biological Invasions, 17*(1), 13–22.
36. Melero, Y., Palazón, S., & Lambin, Y. (2014). Invasive crayfish reduce food limitation of alien American mink and increase their resilience to control. *Oecologia, 174*, 427–434.
37. Lodé, T., Cormier, J. P., & Le Jacques, D. (2001). Decline in endangered species as an indication of anthropic pressures: The case of European mink *Mustela lutreola* western population. *Environmental Management, 28*, 727–735.
38. Podra, M., Gómez, A., & Palazón, S. (2013). Do American mink kill European mink? Cautionary message for future recovery efforts. *European Journal of Wildlife Research, 59*(3), 431–440.
39. Sidorovich, N. V., & Polozov, A. (2002). Partial eradication of the American mink *Mustela vison* as a way to maintain the declining population of the European mink *Mustela lutreola* in a continental area. A case study in the Lovat River head, NE Belarus. *Small Carnivore Conservation, 26*, 12–14.
40. Ahola, M., Nordström, M., Banks, P., Laanetu, N., & Korpimäki, E. (2006). Alien mink predation induces prolonged declines in archipelago amphibians. *Proceedings of the Royal Society B: Biological Sciences, 273*, 1261–1265.
41. Hammershøj, M., Thomsen, E. A., & Madsen, A. B. (2004). Diet of freeranging American mink and European polecat in Denmark. *Acta Theriologica, 49*, 337–347.
42. Zschille, J., Heidecke, D., & Stubbe, M. (2004). Verbreitung und ökologie des minks—*Mustela vison*—In Sachsen-Anhalt. *Hercynia, 37*, 103–126.
43. Salo, P., Nordstrom, M., Thomson, R. L., & Korpimaki, E. (2008). Risk induced by a native top predator reduces alien mink movements. *Journal of Animal Ecology, 77*, 1092–1098.
44. Retrieved from http://www.miljodirektoratet.no/old/dirnat/attachment/2264/DN-utredning-6-2011_nett.pdf
45. Ozolins, J., & Piläts, V. (1995). Distribution and status of small and medium-sized carnivores in Latvia. *Annales Zoologici Fennici, 32*, 21–29.

46. Harrison, M. D. K., & Symes, R. G. (1989). Economic damage by feral American mink (*Mustela vison*) in England and Wales. In R. J. Putman (Ed.), *Mammals pests* (pp. 242–250). London: Chapman and Hall.
47. Retrieved from http://old.furcommission.com/resource/Resources/MFIUS.pdf
48. Retrieved from http://www.furcommission.com/farming/mink-biology/

Marine (Cane) Toad on Pacific Islands

Photo from a-z-animals.com [1]

A Marine Toad That Is Not "Marine" at All

A look at the pattern of distribution of the marine toad over the Pacific islands, including the largest one, the Australian continent (where it is called "cane toad"), combined with the fact that Carl von Linné in his 1758 description of the species named it *Bufo marinus*, may lead to the impression that this is an amphibian that can, on its own, traverse huge oceanic distances. Lately, it was reclassified and given the binomial name *Rhinella marina*.

Linné had gotten the idea that the toad was also a marine animal, from an illustration by a Dutch zoologist, Albertus Seba, who mistakenly believed that the marine toad inhabited both terrestrial and marine environments [2].

A. Jernelöv, *The Long-Term Fate of Invasive Species*,
DOI 10.1007/978-3-319-55396-2_14

That idea, however, was totally wrong. Just as all other amphibians, the marine toad does not live long in seawater, and despite its Latin and popular names, the marine toad is one of the most terrestrial of toads.

The eggs, for sure, are laid in freshwater or brackish water, some 10,000 or even twice that number per female [3, 4]. They do not grow very large as algae-eating larvae and metamorphose at a size of only a cm and a weight that is a mere 0.01% of their eventual mass as adults. The largest recorded specimen of marine toad had a weight of 2.65 kg and a length from the nose to the vent of 38 cm. In nature, though, the adults seldom weigh a kg and mostly clearly less than so, the females being significantly larger than the males. Nevertheless, for toads, that is a lot! Hence, their third common popular name is "giant toad." Initially, as adolescents, they remain near water, as they are dependent on having wet skin, to breathe. Later, however, they become fairly water independent. They eat voraciously and grow quickly. Initially their prey consists mostly of insects, but later rodents, lizards, and everything noticeably smaller than themselves that moves—and many things that do not—become fair game to be consumed.

So the reason why they are present on so many Pacific islands is not their ability to cross the ocean on their own, but that man introduced them there and that they had strong invasive qualities. The reason why man kept introducing them had to do with the experience of an early introduction to a Caribbean island—Puerto Rico.

Marine Toads in Puerto Rico: A Success Story?

In the late nineteenth and early twentieth centuries, the white grubs - larvae of the May beetle (*Phyllophaga* spp.), a scarab beetle - were a major sugarcane pest in Puerto Rico, as well as in many other places. In 1919, marine toads were introduced to Puerto Rico, where they thrived and spread from the coast to inland areas [5]. During this time, the root-eating beetle larvae drastically decreased and became less and less of a problem for the sugarcane growers. The introduction of the marine toad was thus seen as having been a biological control success story and was reported as such at the annual meeting of International Sugar Cane Technologists in Puerto Rico in 1932 [6]. A subsequent article in *Nature* in 1934 with the heading, "Toads Save Sugar Crop," reinforced the message [7]. The abstract of the article reads:

> "Biological control seldom extends to the importation of Amphibia, but great success has followed the establishment of the large toad *Bufo marinus* in Puerto Rico. From two lots of this species brought to Puerto Rico from Barbados and Jamaica, millions of descendants have sprung, and the food of this host has consisted largely of the May beetle (Science Service, Washington, D.C.). The sugar crop, which is the staple product of the island, was threatened by great numbers of the 'white grubs' of May-beetles, which swarmed everywhere in the soil, devouring the roots of the cane and of other plants as well, so that the planters were reduced to picking the grubs by hand. The introduction of the toad has reduced the May-beetles to scarcity, and the Puerto Rican sugar crop has been freed from its worst enemy."

This inspired the introduction of the marine toad throughout the Pacific basin [8] and to more islands in the Caribbean.

Another sugarcane pest, the sugarcane rhinoceros beetle (*Strategus barbigerus*), also disappeared from Puerto Rico after the introduction of *R. marinus*, but remained a pest on a nearby island where the toads had not been introduced [9]. During the 20 years from the early 1920s to the early 1940s, in which the marine toad was very common in Puerto Rico, many species on which it preyed became rare bordering on absent, among them the millipede, *Trigoniulus lumbricinus,* and the Puerto Rican click beetle, the cucubano (*Pyrophorus luminosus*) [10]. As the cucubano larvae are a voracious predator of the white grubs, in this case the new predator almost eliminated an old, but did overall do a better job in protecting the sugarcane.

Unfortunately, Puerto Rico's success in controlling the white grub by introducing the marine toad was not repeated everywhere, so in hindsight it has been argued that demise of the white grub in the cane fields could have had other causes, such as unusual weather conditions that might have prevented the emergence of the beetle pupae. Critics note that the 6-year period, starting in 1931, during which the explosive of marine toad expansive and the dramatic decline of the white grub took place, also saw the highest ever rainfall on the island [8].

When in the mid-1940s the marine toad population on Puerto Rico declined after its initial explosion and population densities dropped to just a small percentage of its peak value, the suppressed insects and millipedes returned in numbers.

The Marine Toad, Its Original Distribution and Natural Enemies

The native homeland of the marine toad stretches from Sonora to Tamaulipas in northern Mexico, including a small part of southernmost Texas, to the river basins of the Orinoco and the Amazon in South America.

The cane toad is an ancient species. A fossil toad from the late Miocene, some 15 to 5 million years ago, from La Venta in Colombia is indistinguishable from modern cane toads from northern South America [11]. The fossil was discovered in a floodplain deposit, which suggests the *R. marina* habitat preferences have always been for open areas [12].

In general, the density of the marine toad is higher in open areas than in thickly vegetated ones, with the exception, of course, of deserts and mountain ranges. In South America, a typical figure along an open shoreline might be 20 toads per 100 m [13] or 30 per 100 × 100 m (= ha), taking the lower density further from the shoreline into account. This is a lot for a large amphibian, but still a small number compared to the density found in the first phase after introduction in many places to which the marine toad has been introduced.

There are many reports claiming that in their native range, several species prey on the marine toad as eggs, larvae, or adults, despite its toxicity. Among those predators are caimans, for example, the broad-snouted caiman (*Caiman latirostris)*; snakes, including the common garter snake (*Thamnophis sirtalis*) and the banded cat-eyed snake (*Leptodeira annulata*); rats; fish, such as the rock flagtail (*Kuhlia rupestris*), several species of killifish, catfish of the Siluriformes order, and eels of the Anguillidae family; opossums (*Didelphis marsupialis*); and several birds, for instance, several species of ibis [1, 14]. Some of the predators have an evolutionary resistance to bufotenine, the main toxin of the marine toad. Others have an inherited eating habit that involves them just making a small hole in the skin to suck blood, and yet others instinctively turn the toad over, slit it open, and feast on the intestines, thereby avoiding the poisonous skin and glands on its back. Some predators, such as crows in areas that the marine toad has invaded, have also learned the latter trick.

Most vulnerable to predation are the young toads immediately after metamorphosis, as their skin glands holding the poison have not yet fully developed [13]. Besides the vertebrate predators mentioned above, ants, centipedes, and wolf spiders may also eat the very young toads. Later, when the toads are considerably larger, the roles reverse and those invertebrates all become possible prey for the amphibian.

By far the most dangerous predator of marine toad tadpoles, however, are older cohorts of the same species [15]. Studies of larvae survival from egg to metamorphosis show high survival rates (60–80%) in the absence of older relatives, but very low survival (1–3%) in their presence. Similarly, the most likely predator of a newly metamorphosed adolescent marine toad is a larger one.

In its native range in South America, the presence of debilitating ticks (*Amblyomma dissimile* and *A. rotundatum*) reduces survival and fecundity of adult marine toads, having a large impact on toad numbers [16, 17].

Toxicity and Defensive Strategies

The marine toad is toxic in all of its life stages. In the eggs themselves, the female deposits high concentrations of bufadienolides—cardioactive steroids. To date, 28 different toxins in the family have been identified, 17 of which have not been found in any other ontogenetic stage [18]. Total bufadienolide concentrations are also highest in eggs (2.64 ± 0.56 μmol/mg). The tadpoles are not able to synthesize their own toxins and concentrations drop, both as a function of selective breakdown with time (fewer types of steroids) and dilution in more body tissue as they grow, until just before metamorphosis they reach a concentration level of only a few percent of what they had in the egg. As adolescents, they gradually develop the ability to produce their own toxins and soon reach high total concentrations, but not until the large parotoid glands behind the eyes are fully developed can they synthesize the full register. Among the toxins they then produce is bufotenin, a tryptamine related to the neurotransmitter serotonin. It has strong hallucinogenic properties, much like heroin and cannabis, and is sometimes used as a cheap alternative to those drugs.

It is therefore listed as a class 1 drug under, for example, Australian and Californian laws.

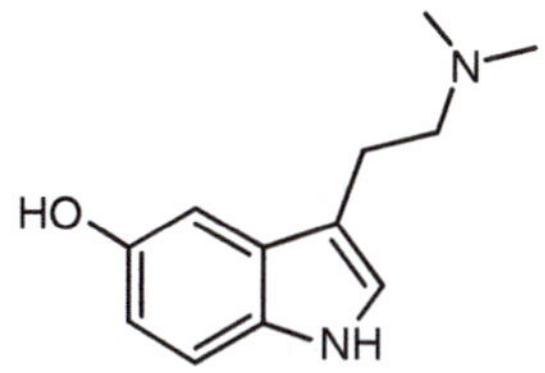

Bufotenin and a co-occurring sister substance with similar effects are the reason why "toad licking" is being used by some drug addicts as a dangerous, but cheap, way to "score" a trip or kick

The mixture of toxins of an adult marine toad is discharged from the glands as a sticky, milky white fluid. When especially threatened, the toad can even forcefully expel the toxins and hit an opponent at a distance of some 30 cm. Mostly, though, the marine toad will only respond to a threat by inflating itself to look more impressive. When actually attacked, it mostly does not respond in any way other than to remain still, perhaps lowering its head better to expose the glands behind its ears, and ooze protective slime all over itself. It then simply waits for the toxins to do their job, and in most cases, they do—and quickly. Attacking dogs, cats, and rats have been seen to become incoherent in less than a minute after contact with the poison and to die within 10 min. In some cases, when the attacker is small, such as puppies or kittens, and the toad is large, it has been observed to eat the dying attacker.

Sometimes the butotoxin secretion can be so voluminous as to practically cover the toad:

Photo from http://www.ntnews.com.au/article/2008/06/10/4342_ntnews.html Visited 16/10 2014

Generally, however, the marine toad is not a nervous animal and is not easily frightened. If dealt with calmly and softly, it can be handled with bare hands without risk.

Some Notes on the Biology and Habits of the Marine Toad

The female lays her large assembly of small (less than 2 mm), black eggs in a long string of jelly. She prefers bodies of water that are calm and warm, which mostly means small, shallow, and nonpermanent. She attaches one end of the string to aquatic plants or other solid underwater objects, and as she swims away the string is drawn out after her, sometimes measuring up to 20 m long. The number of eggs per meter of slime string is about a 1000. As a consequence, the tadpoles often are found in huge swarms, with little available feed in the form of attached algae for them to scrape off of underwater surfaces. They therefore readily accept any additional food, be it a dead animal or terrestrial plants that happen to fall into the water or other, smaller, tadpoles.

When the adolescents appear on land after metamorphosis, at the age of about a month, they are small and not very well developed. Their lungs, for example, have not yet the capacity to supply them with sufficient oxygen, so they depend on supplementary uptake through their skin, which must therefore be kept moist, and their as-yet immature toxin glands only provide them with very limited toxic protection. They have a ravenous appetite, however, and within a year they stretch ten times in all three dimensions and increase their weight a 1000-fold. As they grow, their organs develop to full capacity, and as their skin gets thicker, they become less and less water dependent. As adults, they have an astonishing ability, for an amphibian, to tolerate long dry spells. They survive by losing up to half of their body weight to dehydration [19] and have a number of behavioral tricks, such as digging themselves down into the ground, to reduce water losses [20, 21].

The Marine Toad Here, There, Everywhere

The Puerto Rico success story, with the marine toad as a biological control agent, justified its further introduction to a number of islands with sugarcane plantations, in both the Caribbean and the Pacific, as well as into Florida. However, well before that, in the 1840s, it had already been introduced into Martinique and Barbados [22] for sugarcane pest control and to Jamaica [23] in an attempt to control rats, all with less-than-convincing results.

Hawaii

The marine toad was introduced to Oahu, Hawaii, in 1932 from Puerto Rico, to combat sugarcane insect pests. Within a year and a half, the originally 150 individuals had increased to over 100,000 [24]. Initially, the introduction was seen as another great success, and from the Oahu population, toads were quickly sent to the other Hawaiian islands and many other places in the Pacific region. Altogether, some 600,000 toads and toadlets were "exported" [22].

Today, the toad is very common on Hawaii, especially in suburban gardens around Honolulu. There are hardly any studies of its ecological and economic effects; a large proportion of available articles deal with its toxic effects on household pets such as dogs and cats. In addition, the marine toad itself is sometimes kept as a garden or indoor pet, and handling and feeding recommendations are abundant on the Internet.

Philippines

To the Philippines, the marine toad first came in 1934, when the Philippine Bureau of Plant Industry (BPI) imported 50 specimens from Hawaii. The purpose was to control sugarcane insect pests [25]. A breeding facility was established in Manila at the campus of the University of the Philippines, and the offspring of the imported animals were later released in the sugarcane fields on the island of Negros in 1935 and 1936. In the process of breeding and transport, some toads escaped confinement.

The success of the marine toad in controlling sugarcane pests in Puerto Rico and Hawaii was not repeated in Negros, but the toad nevertheless quickly became established on the island and on Luzon and by 1941 was already common there. In 1954, it was reported also on Panay [26]. Subsequently, through introductions, the marine toad spread to more and more islands in the Philippine archipelago.

Today, the marine toad has colonized most islands across the Philippines, with the Palawan islands being the main exception. There, the native bufonid species of toad (*Bufo philippinicus*) may have prevented *B. marinus* from getting a foothold.

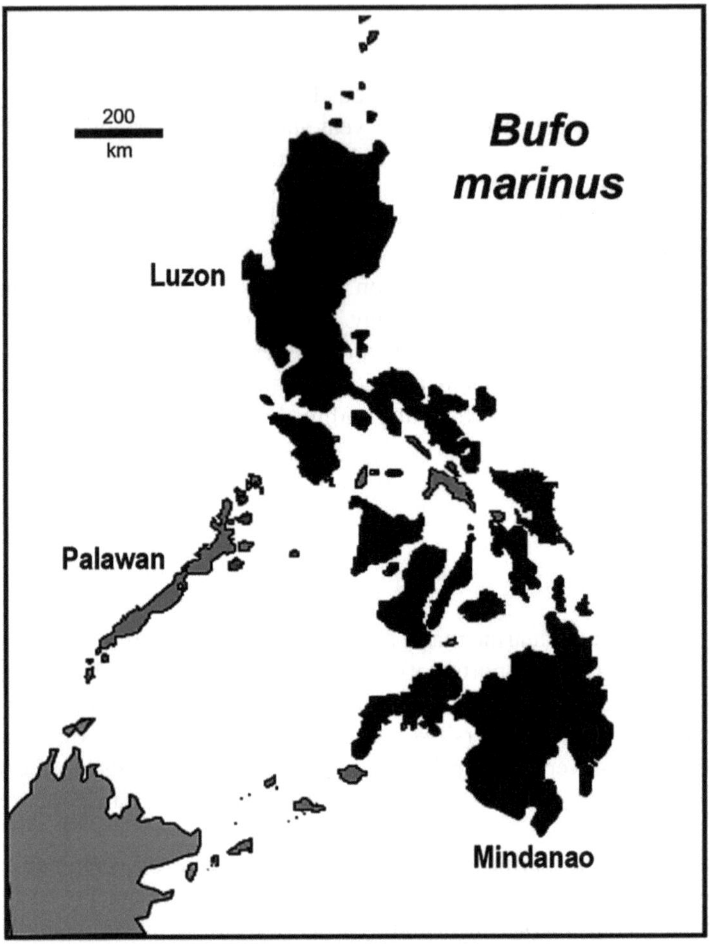

(Dark shaded areas represent the range of the species)

Distribution of *Bufo marinus* in the Philippines [27] (Dark-shaded areas represent the range of the species)

Marine toad thrives especially in human-modified environments and artificial habitats, particularly at low elevations. These habitats include second-growth vegetation, scrubland, forest plantations, cultivated fields, and built-up areas. Although some native frog and toad species coexist with them there, by using degraded rather than natural habitats, the invader mostly occupies a hitherto unclaimed niche.

Studies of population densities along streams in San Mariano on Luzon in 2006 show population densities of *B. marinus* of close to 2000 individuals per ha, while on flooded rice paddies in Los Baños, in 1996, the figure was just over 1000.

Dietary studies showed that 80% of what the marine toad consumed consisted of insects (crickets, ants, beetles, termites, and cockroaches), 13% was earthworms, and 7% was land and freshwater snails [28].

Earlier reports from Negros noted that not only invertebrates were consumed but also blind snakes (probably *Ramphotyphlops braminus*) and house rats (*Rattus tanezumi*) [29]. Likewise, Villadolid in a 1956 study found significant numbers of mice and rats in toad stomachs in the Philippine islands [30].

It is now 80 years since the marine toad was introduced to the Philippines. During that time, only four studies of its ecological impact have been carried out and published. Noteworthy is that they all concluded that the marine toad has had limited impact on native amphibians, due to dissimilar habitat preferences with the invader in disturbed and anthropogenic habitats [27–30].

In Philippine English, the marine toad is commonly known as "bullfrog" and in Visayan languages it is called "kamprag" meaning "American frog."

Fiji

The decision to introduce the marine toad to Fiji was a slow and drawn-out process. It was first initiated in 1933 following the successes in Puerto Rico and Hawaii. Concerns of possible negative side-effects were raised, however, and kept popping up until after 20 years, in 1953, the national government took the decision to import 67 specimens from Hawaii and release them to combat insects on infested sugarcane fields [31]. A decade later, in 1963, when the marine toad was firmly established on the major islands in Fiji, an outcome study was undertaken [32]. It concluded that the toads' diet had included both harmful and beneficial invertebrates, that insect damage to sugarcane had been reduced but not eliminated, and that, unexpectedly, household cockroaches had become noticeably fewer. All in all, the impact of the marine toad was considered "economically neutral"—an assessment that both the pro- and anti-toad lobbies could live with. A peculiarity noted with regard to the marine toad on Fiji is its relatively small size, seldom exceeding 6–7 cm in length and half a kg in weight [31].

New Guinea

In the 1930s, the larvae of the hawk moth (*Hippotion celerio*) were found to be severely damaging the sweet potato crop in New Guinea, and in 1937 marine toads imported from Hawaii were released to combat the pest [33]. Further releases followed, with "wild" toads from Australia and "laboratory" ones that had been used for pregnancy tests. Initial reports found that the marine toad had indeed been effective against the so-called cutworm and that sweet potato yields had improved [31]. Following that, the toads were deployed across much of the region to protect all sorts of agricultural plants.

To New Britain Island in the Bismarck Archipelago of Papua New Guinea, the marine toad was introduced in 1939. Here, besides sweet potato, cacao was becoming an important crop, but it suffered being set upon by the cacao weevil borer

(*Pantorhytes plutus*). In a study of the food habits of marine toad living on and in the litter layer beneath the cacao trees and the canopy of shade trees, it was found that ants constituted a large part of their diet (46%) followed by snails (42%). Adult cacao weevil borers were present in the toad stomachs but constituted only some few percent of the content [34].

As weevils and the snails damage the cacao plants and fruit, while ants help to protect them, it remains unclear what the presence of the marine toad means for cacao production. What is clear, however, is that the toad's diet reflects the relative abundance of food organisms. Other studies found larger proportions and numbers of cacao pests eaten by *Bufo marinus* in New Guinea [35].

Guam, Palau, and Marshall Islands

The story of marine toad introduction on these islands is a different one from those of most other islands. Here, another invasive species had been introduced first: the monitor lizard (*Varanus indicus*). In some cases, the Polynesians were responsible, probably as they wanted to have the big lizard as a source of meat. In other cases, the Japanese brought the monitor lizard to control rat populations. Regardless of the timing of, and the reason for, their introduction, the lizards showed an appetite for chicken and subsequently prevented the establishment of, or eliminated the existing, chicken agro-industry.

With the American entry into World War II and arrival to the Pacific islands came an attempt to resolve the monitor lizard problem—by introducing the marine toad, whose toxins would kill the offensive lizard. The control attempt was at least partly successful or expressed in a different way: "Toads have been implicated in the decline of the population of monitor lizards in Guam" [36].

Since then, both the lizard and the toad have been overtaken as the invader of most concern by the brown tree snake (*Boiga irregularis*).

Japan

The first place in Japan to which the marine toad was introduced was Minamidaitojima archipelago. It took place in the late 1930s, and toads from Hawaii were brought there for sugarcane beetle pest control. The outcome was seen as successful, and in 1949 specimens from Minami were taken to Chichijima, in 1974 to Hahajima, and in 1978 to Ishigakijima. The spread to Hatomajima in 1984, however, seems to have been accidental; the toad probably hitchhiked on a load of building material.

The toad was found to reduce ground invertebrates through predation, amphibians and small reptiles through competition, and some large predators through toxicity. One aspect of the latter, namely, the reduction of feral cat populations, was seen as positive. A specific negative effect was the poisoning of scarce freshwater by toxic tadpoles [33].

Australia

In 1935, some 3000 marine toads from Hawaii were released in the cane fields of New Queensland for the purpose of beetle pest control. Once in Australia, the toads were quickly renamed "cane toads" and are still known under that name today even though they did not do a particularly good job as sugarcane field wardens. However, they quickly became established and flourished in the new environment [37], steadily expanding their range through Queensland. By 1978, they had reached New South Wales and by 1984 the Northern Territory. It was expected that when they reached drier areas, their spread would slow down from their early average of 40 km per year [38], but instead it has increased, to some 60 km per year on the western front [39].

An explanation was found in that toads at the forefront of the spread had longer legs (and could thus move faster), as a consequence of genetic selection.

In Australia, the cane toad is currently seen as a "key threatening species" and is also officially identified as such in the country's Biodiversity Conservation Act. Long-term negative ecological effects, however, are as yet not well documented and are mostly expressed in general phrases such as "the depletion of native species that die eating cane toads; the poisoning of pets and humans; depletion of native fauna preyed-upon by cane toads; and reduced prey populations for native insectivores, such as skinks" [40].

Some reports, though, are more specific. Populations of northern quoll (*Dasyurus hallucatus*) have shown severe declines after toad invasions of their habitats and so have the populations of several species of Australian monitor lizards and snakes [41], one example being a local population of the Argus monitor (*Varanus panoptes*) that dropped by 90% [42].

Freshwater crocodiles (*Crocodylus johnsoni*) have also been found to be negatively affected by invading cane toad [43], although for saltwater crocodiles (*Crocodylus porosus*), the toads may positively influence the hatching rate by reducing the populations of, inter alia, monitor lizards, that feed on their eggs. There are some conflicting reports as to the sensitivity of saltwater crocodiles to the toad's toxins. Some reports talk about specimens being found dead after ingesting cane toads, while others highlight cases in which *Crocodylus porosus* could both mouth and ingest cane toads with apparent impunity [44].

A crocodile poisoned by a cane toad (Photo: Alicia Bridges, West Australian, 4 March 2011)

With regard to one of the lizard species that was severely affected by the cane toad in Queensland, the blue-tongued lizard, it was found that populations found in other parts of Australia, for example, the Sydney area to which the cane toad has not (yet) spread, were tolerant to the toxin. It turned out that a toxic invasive plant known as the "mother of millions" (*Bryophyllum delagoense*) from Madagascar contains a poison much like the one the cane toad manufactures. The blue-tongued lizards that live in areas in which the plant exists, have developed a tolerance to its toxin. They are thereby preadapted to deal with the toad they have yet to meet [45, 46].

Other species tolerant to the cane toad toxins include the saw-shelled turtle (*Myuchelys latisternum*), which lives along rivers and streams in New South Wales [47], and the keelback snake (*Amphiesma mairii*) [8].

Meat ants (*Iridomyrmex purpureus*) are also immune to the toad toxins, and they exploit this advantage by preying on young cane toads that do not have the instinct to try to escape the ants, but instead remain still when attacked, waiting in vain for their toxins to save them [48].

Without being immune, some species of Australian birds such as the Torresian crow (*Corvus orru ceciliae*) and the black kite (*Milvus migrans*) have learned how to eat cane toads without exposing themselves to the toxin. They flip the toad over, quickly rip-open its belly, and consume the toxin-free inner organs. The owllike tawny frogmouth (*Podargus strigoides*) consumes cane toads [8] using the same technique.

As mentioned earlier the spread of the cane toad in Australia continues and at an increasing speed. Typically, the population in a newly invaded area goes through a transitory population boom with densities as high as one toad per 10 m^2, near water. Later, the numbers drop to a tenth of that or less, with occasional flare-ups [49, 50].

Distribution of the cane toad in Australia 2007. Today the northern part of Western Australia has also been invaded. Picture from Wikipedia. http://en.wikipedia.org/wiki/File:Bufo_marinus_australian_range.png

Ecological and Economic Aspects of Marine Toad Invasions

During the population boom that more or less regularly follows its introduction, the biomass alone of the voracious marine toad, which often amounts to several 100 kg/ha, clearly indicates that it will have a strong impact on the populations of all of its prey. However, the depressed populations of these mostly bounce back after the marine toads have passed their peak, and their numbers go down an order of magnitude or more.

Mostly, and certainly in Australia, the kill-off of predators trying to eat the toxic toad and the subsequent decline of the predator populations is seen as a worse problem, given cascading effects in the ecosystem and the slow recovery of the population of the often large, long-lived, low-fertility species such as dingoes, blue-tongued lizards, and crocodiles [51], but there are no long-term studies to substantiate or refute those concerns.

There have also been very few serious studies of the economic aspects of the marine toad. Most authors who have addressed these aspects at all conclude that there often are positive effects with regard to pest control, but also negative ones, and without much rigor in quantification, they come out with conclusions similar to the one from the study in Fiji mentioned earlier, saying that "The marine toad is economically-neutral."

In this respect, though, Australia may differ from smaller Pacific islands in that the toad was unable to control the cane pest there so its positive effects are truly minor, whereas the problems following its large-scale spread are more pronounced and costly.

Eradication and Control Attempts

Generally speaking, there have been very few attempts to eradicate the marine toad, as they were as a rule deliberately introduced and their initial establishment was welcome. Later, when sentiments may have shifted, it was simply impractical. Chemical control, for instance, the poisoning of bodies of water in which the tadpoles live, has been tried with limited success and with a heavy cost in the form of damage to native amphibians.

Physical removal has been and is often practiced. With adult toads, they are simply picked up and disposed of in any which way. Local toad populations can temporarily be reduced in this way, but enthusiasm for this nocturnal activity among, for instance, suburban residents or environmental groups, tends to wear off rather quickly. A number of toad traps can also be found on the market, some with sophisticated olfactory or acoustic attractants.

With tadpoles, which often aggregate in huge numbers, netting campaigns are easier to sustain for the limited time periods in question. Some attempts have also been made to increase the catches by concentrating the tadpoles further, by adding toad toxins to a specific pond site. The tadpoles are attracted by the toxins, as they presumably associate them with the presence of new eggs or newly hatched larvae, on both of which they happily cannibalized. Attractant pheromones have also been used for the same purpose. The netted tadpoles are then often killed by immersing them in hot water or by simply dumping them on dry ground. With this method, however, the toad population can only be noticeably reduced when there are few water bodies available for spawning.

That leaves biological control options, of which a number have been studied, especially in Australia. Alarm pheromones create panic among tadpoles, and repeated applications can stress them to death or delay their development by reducing their appetite [52].

The parasitic lungworm (*Rhabdias* sp.), naturally occurring and found among toads in the Northern Territory and Western Australia, was seen as a prime candidate around 2010, but did not live up to the hopes for its efficacy.

A research project at CSIRO also set out to construct recombinant viruses working on genes with different expressions in tadpoles and toadlets. However, the project was not finalized to execution due to concerns about the release of genetically modified organisms (GMOs) [53].

Attempts have also been made with the release of small sterile males ahead of the advancing invasion front, to "teach" predators in a nonlethal way to avoid eating cane toads and with marine toad males who will only get male offspring to reduce overall toad fertility. Also these trials were interrupted before any large-scale applications took place. Thus, so far, no workable biological control method for the cane toad has been found [54].

The Marine Toad and the Human Environment

Also in its native area of distribution, the marine toad thrives best on open or semi-open land and has clearly lower population densities in dense forests. Often, these preferred areas are the results of human land clearings for agricultural or logistical purposes (such as road construction).

In the areas of introduction, this pattern is even more obvious. Not only were humans the vehicles that brought the marine toad to these often faraway places, but they also put the toads where they wanted them, which was mostly on sugarcane fields. However, when the toads became established, they spread to all available habitats, with those disturbed by man (their favorites), supporting the highest population densities.

Over the last several decades, the marine toad has taken further steps to become man's follower. Suburban neighborhoods with sprawling gardens have become fertile new hunting grounds, not only because gardens harbor many edible organisms, for example, in compost piles, but also because the marine toad has learned to eat food set out for pets such as dogs and cats and to ransack garbage bags and dumping sites. In addition, people often deliberately feed them.

Living in the vicinity of man also carries certain risks, however. Marine toads often have a penchant for asphalt roads, and their strategy for defense against natural predators (sitting still and trusting their toxins) does little to protect them from motorized vehicles. Thus, being run over has now become a major cause of death for the marine toad.

Human Utilization of Marine Toads

The arguably oldest recorded human use for the marine toad was the application of their toxins to arrow tips for hunting, utilized by Indians in South and Middle America in much the same way as they did with the toxins from poison arrow frogs of the Dendrobatidae family. Whether they also smoked the foamy slime to intoxicate themselves, as some modern drug users claim, is more questionable.

The modern use of marine toad toxins as drugs has been mentioned above. In Japan, the bufotenine toxin has been used as an aphrodisiac and as a hair restorer, and as a dried powder, it has found its way into the Chan Su version of traditional

Chinese medicine, both as a topical anesthetic and as an aphrodisiac. Furthermore, it is used to lower the heart rate of patients undergoing cardiac surgery, both in Taiwan and on the Chinese mainland.

Currently studies are under way of marine toad poison as a treatment for prostate cancer [55]. During some decades in the middle of the twentieth century, the marine toad and several other frog and toad species were used for human pregnancy tests. The urine of women was injected into lymph glands of male amphibians, and if the woman was pregnant, spermatozoa appeared in the animal's urine [31].

In many places, perhaps particularly in the Philippines, marine toad leather is used to make a number of products, including purses, valets, handbags, gloves, hats, and vests. In addition, stuffed toads are big items on the curio market.

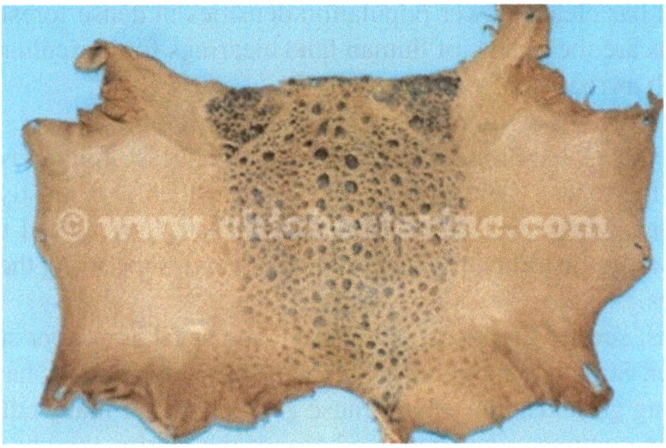

A piece of cane toad leather. The skins are 10–12 × 12–15 cm and cost $20 (Photo: Chichester Inc. Philippines)

The marine toad itself also serves as a tourist attraction. In many places, both where the toad is native and where it has been introduced, nocturnal safaris are organized to look for them in situ. One such place is Drake Bay, in Costa Rica, where organizers tell stories about a 12 cm toad caught on camera swallowing a 30 cm sock-headed snake (*Enuliophis sclateri*) [56].

The marine toad has also found a place and a market as a pet. The largest recorded specimen was kept as such in Sweden and was named "Prinsen" (The Prince). It was 38 cm from the snout to the vent and weighted 2.65 kg.

Summary

The marine toad has been introduced to a large number of islands in the Caribbean and the Pacific, including the Australian continent. Mostly, these introductions have led to the establishment of invasive populations with very high densities during their initial boom, which mostly lasts for several decades.

Especially during this initial period of high toad population density, the effects on local ground invertebrates, amphibians, and reptiles have been pronounced. Many predator species have also suffered significant losses due to exposure to toad toxins. Later, when the toad populations have dropped, the effects on native and other invasive species have become less severe. So far, no extinction of species, even locally, has been attributed to the introduction of the marine toad.

Practically all of the Pacific islands, on which introduction mostly took place in the 1930s, have entered the second, more calm phase of marine toad presence. Australia is in a way a special case, as the toad is still spreading over the continent, with the most recently conquered fronts still being in the first, intensive phase of invasion.

Anywhere in the Pacific region in which the marine toad established a foothold after its introduction, it now seems firmly entrenched with little probability of it being dislodged or exterminated.

References

1. Retrieved October 15, 2014, from http://a-z-animals.com/animals/marine-toad/
2. Beltz, E. (2006). *Scientific and common names of the reptiles and amphibians of North America—Explained*. Boston: Houghton Mifflin.
3. Oliver, J. A. (1949). The peripatetic toad. *Natural History, 58*, 30–33.
4. Bartlett, R. D., & Bartlett, P. P. (1999). *A field guide to Florida reptiles and amphibians*. Houston, TX: Gulf Publishing.
5. Grant, C. (1931). Notes on *Bufo marinus*. *Copeia, 1931*, 62–63.
6. Tyler, M. J. (1989). *Australian frogs*. Victoria, Australia: Penguin Books.
7. Toads Save Sugar Crop. (1934). *Nature, 134*, 877.
8. Freeland, W. J. (1985). The need to control cane toads. *Search, 16*(7–8), 211–215.
9. Wolcott, G. N. (1950). The sugar cane rhinoceros beetle. *Journal of Economic Entomology, 43*, 385.
10. Wolcott, G. N. (1948). What has happened to the giant Surinam toad, *Bufo marinus* L. in Puerto Rico? *Revista de Agricultura de Puerto Rico, 38*, 94–98.
11. Estes, R., & Wassersurg, R. (1963). A Miocene toad from Colombia, South America. *Breviora, 193*, 1–13.
12. Zug, G. R., & Zug, P. B. (1979). The marine toad, *Bufo marinus*: A natural history resume of native populations. *Smithsonian Contributions to Zoology, 284*, 58.
13. Lampo, M., Leo, D., & Giulio, A. (1998). The invasion ecology of the toad *Bufo marinus* from South America to Australia. *Ecological Applications, 8*(2).
14. Cohen, M. P., & Alford, R. A. (1993). Growth, survival and activity patterns of recently metamorphosed *Bufo marinus*. *Wildlife Research, 20*, 1–13.
15. Alford, R. A., Lampo, M., & Bayliss, P. (1995). *The comparative ecology of Bufo marinus in Australia and South America*. CSIRO Bufo Project: An Overview of Research Outcomes, CSIRO, Australia.
16. Lampo, M. (1995). *The identification of ticks that infect Bufo marinus in South America and their impact on toad populations*. CSIRO Bufo Project: An overview of research outcomes, Unpublished report, CSIRO, Australia, Centro de Ecologia, Venezuela.
17. Retrieved October 17, 2014, from http://amphibiaweb.org/cgi/amphib_query?where-genus=Rhinella&where-species=marina

18. Hayes, R. A., Crossland, M. R., Hagman, M., Capon, R. J., & Shine, R. (2009). Ontogenic variation in the chemical defenses of cane toads (*Bufo marinus*): Toxin profiles and effects on predators. *Journal of Chemical Ecology, 35*(4), 391–399.
19. Krakauer, T. (1970). Tolerance limits of the toad, *Bufo marinus*, in South Florida. *Comparative Biochemistry and Physiology, 33*, 15–26.
20. Straughan, I. R. (1966). The natural history of the "cane toad" in Queensland. *Australian Natural History, 15*, 230–232.
21. Carpenter, C. C., & Gillingham, J. C. (1987). Water hole Fidelity in the marine toad, *Bufo marinus*. *Journal of Herpetology, 21*(2), 158–161.
22. Easteal, S. (1981). The history of introductions of *Bufo marinus* (Amphibia: Anura); a natural experiment in evolution. *Biological Journal of the Linnean Society, 16*(16), 93.
23. Lannoo, M. J. (2005). *Amphibian declines: The conservation status of United States species*. Berkeley, CA: University of California Press.
24. Pemberton, C. E. (1933). Introduction to Hawaii of the tropical American toad *Bufo marinus*. *Hawaiian Planters' Record, 38*, 15–16.
25. Merino, G. (1936). Introduction of the giant toad (*Bufo marinus*) into the Philippines. *Philippine Journal of Agriculture, 7*(2), 283–286.
26. Inger, R. F. (1954). Systematics and zoogeography of Philippine Amphibia. *Fieldiana: Zoology, 33*(4), 183–531.
27. Diesmos, A. C., Diesmos, M. L., & Brown, R. M. (2005). Status and distribution of alien invasive frogs in the Philippines. *Journal of Environmental Science and Management, 9*(2), 41–53.
28. Alcala, A. C. (1957). Philippine notes on the ecology of the giant marine toad. *Silliman Journal, 6*(2), 90–96.
29. Rabor, D. S. (1952). Preliminary notes on the giant toad, *Bufo marinus* (Linn.), in the Philippine Islands. *Copeia, 4*, 281–282.
30. Espiritu, L. M. (1985). *Population Ecology and Distribution of Bufo marinus L. in the Province of Isabela, Philippines*. Master's thesis, University of the Philippines at Los Baños, Laguna.
31. Lever, C. (2001). *The cane toad. The history and ecology of a successful colonist* (230). Otley: Westbury Publishing.
32. Hinckley, A. D. (1963). Diet of the giant toad, *Bufo marinus* (L) in Fiji. *Herpetologica, 18*, 4.
33. Retrieved from http://www.nies.go.jp/biodiversity/invasive/DB/detail/40010e.html
34. Bailey, P. (1976). Food of the marine toad, *Bufo marinus*, and six species of skink in a cacao plantation in New Britain, Papua New Guinea. *Australian Wildlife Research, 3*, 185–188.
35. Pippet, J. R. (1975). The marine toad, *Bufo marinus*, in Papua New Guinea. *Papua New Guinea Agricultural Journal, 26*, 23–30.
36. Retrieved October 21, 2014, from http://www.inriodulce.com/links/bufomarinus.html
37. Retrieved October 30, 2014, from http://animals.nationalgeographic.com/animals/amphibians/cane-toad/
38. Tyler, M. J. (1994). *Australian frogs a natural history* (p. 112). Chatswood: Reed Books.
39. Retrieved October 30, 2014, from http://www.abc.au/news/2014-10-09/cane-toad-advance-genetic-research-rick-shine/5799008
40. Zug, G. R., Lindgrem, E., & Pippet, J. R. (1975). Distribution and ecology of marine toad, *Bufo marinus*, in Papua New Guinea. *Pacific Science, 29*(1).
41. Mayes, P. J., Thompson, G. G., & Withers, P. C. (2005). Diet and foraging behaviour of the semi-aquatic *Varanus mertensi* (Reptilia: Varanidae). *Wildlife Research, 32*, 67–74.
42. Doody, J. S., et al. (2007). A preliminary assessment of the impacts of invasive cane toads (*Bufo marinus*) on three species of Varanid lizards in Australia. *Mertensiella, 16*, 218–227.
43. Letnic, M., Webb, J. K., & Shine, R. (2008). Invasive cane toads (*Bufo marinus*) cause mass mortality of freshwater crocodiles (*Crocodylus johnstoni*) in tropical Australia. *Biological Conservation, 141*, 1773–1782.
44. Covacevich, J., & Archer, M. (1975). The distribution of the cane toad, *Bufo marinus*, in Australia and its effects on indigenous vertebrates. *Memoirs of the Queensland Museum, 17*, 305–310.

45. Retrieved from http://www.ecology.com/2012/03/06/invasive-plant-protects-from-invasive-toad/
46. Price-Rees, S. J., Brown, G. P., & Shine, R. (2012). Interacting impacts of invasive plants and invasive toads on native lizards. *The American Naturalist, 179*(3).
47. Ryan, M., & Editor, G. (2000). *Wildlife of tropical North Queensland: Cooktown to Mackay* (p. 198). Queensland Museum.
48. Ward-Fear, G., Brown, G. P., & Shine, R. (2010). Using a native predator (the meat ant, *Iridomyrmex reburrus*) to reduce the abundance of an invasive species (the cane toad, *Bufo marinus*) in tropical Australia. *Journal of Applied Ecology, 47*(2), 273–280.
49. Retrieved October 30, 2014, from http://www.genetics.org/content/159/4/1671.full
50. Retrieved from http://www.ctahr.hawaii.edu/coqui/documents/RAAlfordCaneToadsAustralia.pdf
51. van Dam, R. A., Walden, D. J., & Begg, G. W. *A preliminary risk assessment of cane toads in Kakadu National Park* (Scientists Report 164). Supervising Scientist, Darwin, NT.
52. Australian Academy of Science. (2008, May 10). Biological weapons to control cane toad invasion in Australia. *ScienceDaily*.
53. Retrieved from http://www.environment.gov.au/system/files/resources/ff131ef0-c86c-4620-bc2e-808b88ee9e3d/files/cane-toad-2008.pdf
54. Shanmuganathan, T., Pallister, J., Doody, S., McCallum, H., Robinson, T., Sheppard, A., Hardy, C., Halliday, D., Venables, D., Voysey, R., Strive, T., Hinds, L., & Hyatt, A. (2010). Biological control of the cane toad in Australia: A review. *Animal Conservation, 13*, 16–23.
55. Retrieved from http://www.abc.net.au/news/2014-09-17/cane-toad-venom-attacks-cancer-cells/5750114
56. Retrieved October 13, 2014, from http://www.thenighttour.com/alien1/marine_toad.htm

Raccoons in Europe (Germany)

Raccoon (*Procyon lotor*) Photo from Planetwissen

How They Got to Europe

Raccoons are native to North and Central America. Most stories circulating about their introduction to Europe involve German zoos and World War II. One of them tells about the raccoons that Hermann Göring, a leading Nazi politician with a flashy lifestyle who as Reichsmarschall was the second in command after Adolf Hitler, kept at his private zoo. There, they lived better during the war than most Germans did, but at the very end of the war food became scarce even for them.

A. Jernelöv, *The Long-Term Fate of Invasive Species*,
DOI 10.1007/978-3-319-55396-2_15

When American troops arrived at the place, the US soldiers, who recognized the animals from back home, felt sorry for them and released them from their cages.

Other stories deal with public zoos, such as those in Berlin, Hamburg, and Köln, and with fur farms that were bombed and damaged and from which raccoons succeeded to escape. Yet other stories tell about both accidental and intentional releases from fur farms in the 1950s and 1960s. One of the most-cited ones tells about the release of two pairs north of Hesse in 1934. A poultry farmer who had kept the raccoons as pets applied for a permit from the Prussian hunting office to release them at the Edersee reservoir "to enrich the fauna." He finally received the permit, but according to the story had already released in advance of receiving it [1]. This story also brings Hermann Göring into the picture, by pointing out that he was the highest in command over the permit-granting body.

There had already been earlier attempts to introduce raccoons in Germany for the same purpose – fauna enrichment – but those animals were never seen again, and the experiments were seen as failures. One such attempt took place in 1927 in the same region north of Hesse as the famous 1934 release [2].

In all likelihood, more than one release succeeded, as by the end of the twentieth century, the German raccoon population had two distinct centers, one north of Hesse and one east of Berlin. The two populations could also be distinguished on the basis of an intestinal worm, *Baylisascaris procyonis*, that was prevalent in the Hesse population but absent from the Berlin one [1].

In Systema Natura, Where Do They Belong?

As mentioned above, raccoons are native to North and Central America. The oldest written record of the species is from Christopher Columbus. Taxonomists had problems in placing the raccoon: Carl von Linné first gave its scientific name *Ursus cauda elongata* (long-tailed bear), but then later renamed it *Ursus lotor* (washer bear) [3].

Besides bears, raccoons have been classified at various times as belonging to the families of dogs, cats, and badgers. Currently, it is placed in a family called Procyonidae (half bears). In that family, the raccoon is the largest member, with a body length of up to 70 cm and a weight of 9 kg. Its black face mask and long tail are characteristics that make the animal unique in appearance and easy to identify. Another distinctive feature is the raccoon's very dexterous front paws.

Photo from Wikipedia

Some Raccoon Facts

The raccoon is a mammalian. After a gestation period of little over two months, the female gives birth to 2–5 “kits.” The kits stay with their mother until the autumn and are then left to fend for themselves, although female offspring can be permitted to stay in the territory of the mother.

An omnivore, the raccoon’s diet is dominated by invertebrates, but with a significant quantity of fruits, nuts, flowers, and leaves, and as many smaller vertebrates as they can get a hold of. They are mostly nocturnal, but if feeding opportunities appear at daylight, they will not forego them.

A behavior that is typical for the raccoon has given it its Latin name, as well as its translation in German, French, and many other languages: If water is to be found anywhere near the food source, the raccoon dips/washes its food in the water before eating it. The French naturalist Georges-Louis Leclerc, in his eighteenth-century *Histoire Naturelle*, hypothesized that the raccoon had insufficient saliva production and therefore had to wet the food before ingesting it. For two centuries this was thought to be the case, but not anymore. Modern researchers have ruled out a lack of saliva, but failed to come up with an alternative explanation for the washing behavior.

The ideal habitat for the raccoon is a mixed forest with hard- and softwood trees, some of them old and with hollows, in the vicinity of lakes or rivers. They use the hollows as daytime sleeping quarters and as dens both during the winter and when raising their litters. Alternatively, if no suitable places above ground can be found, the raccoon will use rock crevices or earth burrows excavated and/or previously used by rabbits or badgers. In cold climates, raccoons hibernate over winter and then often prefer (well-insulated) underground dens to tree hollows. In such cases, young females might share the winter quarters with their mother.

Generally, raccoons are solitary animals, although as noted above, related females can share a territory. Sometimes a group of males, not necessarily related, can also form a band to keep other males out, especially at mating times. Given this general urge for privacy, it may seem a little strange that the raccoons, even when living on their own, often share a "toilet," a place where they come to dispose of urine and feces.

With regard to the senses, the raccoon relies very much on its sense of smell in its social relations and on its sense of touch when searching for food. Its eyesight is not bad, but they are nearsighted. That, together with their being neither especially fast nor good jumpers, means the raccoon is not a good hunter. Many of the vertebrates they consume, such as amphibians or fish, are caught in shallow waters. Eggs and young birds in their nests are among other not-too-difficult-to-catch types of vertebrate prey.

Raccoons are seen as intelligent animals. Much of that fame goes back to studies by H. B. Davis [4], when he tested their ability to open locks. In that, they compared well with rhesus macaques. Later studies, for example, by B. Pohl in 1992 [1], focused on their memories, and in that area they also came out well.

"Demographics" of Raccoons in Their Natural Habitats

The population densities of the raccoon in North America vary with the type of ecosystem and food availability, ranging from 5 to 6 individuals per square kilometer in hardwood forests to from less than one to a few on prairies [1]. Another way of saying this is that the size of the home range for male raccoons on the prairies of North Dakota is between 7 and 50 km^2 (and a third of that for females) and 0.5 km^2 for both sexes in the Lake Erie marshes. Generally speaking, older raccoons, especially males, claim larger territories than younger ones.

In captivity, the animal can live up to 20 years old, but in nature that occurs very seldom, if ever. The average lifespan is only 2–3 years. Roughly about half of the young survive their first year, but after that annual mortality goes down to 10–30% [1].

Man, through hunting and roadkill, is a major cause of death for raccoons [5]. Hunting, however, has decreased significantly in recent decades. Prices for raccoon fur have fallen, not least after it became unpopular to wear fur in, for example, Europe, a major market. The opposite had been the case from the 1940s through the

1970s, when movies and TV series spurred the market and made fur hats and coats much sought-after. From just short of 400,000 raccoons killed in the hunting season of 1934–1935, the seasonal hunt figures followed the population increase and reached one million in 1946–1947 and two million in 1962–1963. The peak was reached in 1976–1977, when 5.2 million raccoons were killed and skinned in the USA alone. Since then, the hunt has fallen sharply to well under a million animals shot and trapped annually.

At certain times and places, disease is also a major cause of raccoon deaths. Most lethal is canine distemper, caused by a single RNA-strand virus closely related to measles. It is thought to be a new animal disease, with the first case described only in 1905, by the French veterinarian Henri Carré [6]. In all likelihood, it was transmitted to North American raccoons from infected domestic dogs.

Raccoons also carry rabies, a viral disease that for them is lethal. In the wild population in North America, its spread started in the 1950s in the southeastern US states of Florida and Georgia.

There are several mammalian predators in North America that may eat raccoons, including wolves, coyotes, bobcats, and lynxes. Great-horned owls and eagle owls may also have them on the menu, but overall, predation is not a major cause of death for the raccoon.

In the last seventy years, raccoons have increased drastically in numbers and conquered much new territory. The most visible result of this is their prevalence in cities, in which their population densities are often much higher than in forests. In the cities, they feed on garbage, fruits from gardens, animal feed, and the eggs and chicks of doves, sparrows, and other urban birds.

The raccoon has also extended its range northward, spreading well into the middle and northern parts of Canada's southern provinces, mountain ranges such as the Rockies, and coastal marshes.

Some estimates say that overall the US raccoon population has expanded 15–20 times between the 1930s and the 1980s [5]. Other estimates talk about a 70-fold increase over a hundred years. This is at least partly supported by state figures, indicating a tripling in two decades from the early 1980s, for example, in Illinois [7].

There are few figures available on the total numbers of raccoons. According to Blitz Enterprises, a company dealing with raccoon damage control, there were some 1.1 million individuals of the species in southern Ontario in 2012. That number agrees fairly well with the population density figures quoted above and the areas of agricultural lands (54,000 km^2) and hardwood forest (100,000 km^2) of the province, given that some cities, especially Toronto, also have a large raccoon population. (More about that later.)

If those raccoon density figures are applied to the whole of North America, with its 3 million km^2 of hardwood forests and 4.8 million km^2 of agricultural land, the calculated size of the population would be 35–40 million. Add in the city populations, and one reaches a total rough estimate of 40 million animals.

Urban Raccoons

A number of animals and birds are making the urban environment, and especially its suburbs, their new home. The raccoon is one of them. Cincinnati was the first city in which raccoons established a population (in a suburb). This happened in the 1920s, and since then urban populations of raccoons have become common all over North America, with major metropolitan areas such as New York, Chicago, Washington, D.C., and Toronto getting their first new inhabitants in the 1950s and seeing successively increasing numbers since then [7]. In the cities, raccoons find shelter and build their dens in attics, abandoned houses, garages, garden cottages, and hollows of old trees in parks. They feed on garbage, garden fruits, and pet food, in addition to, for example, the eggs and chicks of urban birds such as pigeons and sparrows. In the urban environment they are often seen as marauders, denounced as "masked bandits," and cause not insignificant economic damage. Given the large amount of food and shelter available to them in their urban environment, raccoon density is much higher than in rural areas, with up to a hundred individuals per km^2. However, these densities in urban areas are, as yet, far from common. If they were to become the standard, the urban raccoon population in North America could come to number as much as a third of the whole.

The Development of the German Raccoon Population

As noted above, the German population in the mid-1990s had two distinct centers, one north of Hesse and one east of Berlin. Their numbers were then estimated to be close to 100,000, up from some 20,000 two decades earlier and a mere 285 in 1956, another two decades before that.

Their distribution in the first years of the third millennia is mapped below. The raccoons by then had spread to all of Germany and into the neighboring countries. The two original populations had met, but still constituted separated areas with the highest densities.

On the map, a small high-density pocket of raccoon in northern France can also be spotted. This is thought to represent an introduction of its own, carried out in 1966 when members of the US Air Force, facing short-notice transfer to Vietnam, released some pet raccoons outside the Laon-Couvron Air Base [5].

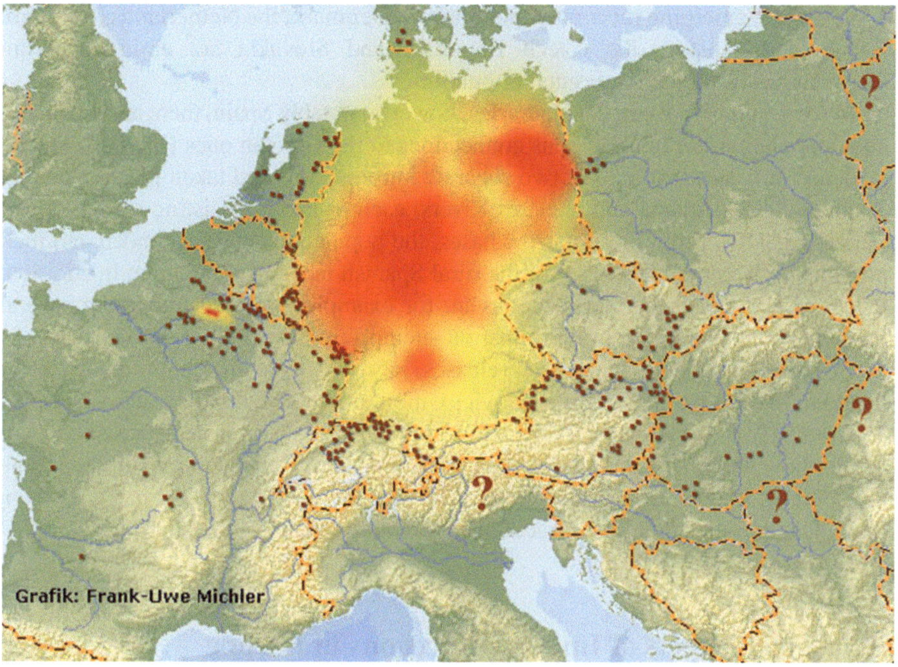

Raccoon distribution in Germany 2000–2003 (from Frank-Uwe Michler)

Since the mid-1990s, the number of raccoons has again more than quadrupled, and conservative estimates made in 2013 put their numbers at over a half a million. Others, who base their estimate of the population increase on figures from the Deutscher Jagdverband (DJV), a hunting association, talk about a million. During the hunting season 2012–2013, over a hundred thousand raccoons (104,371) were killed either by traffic or by hunters, up from 71,071 the season before and from some 30,000 in 2007 and 5000 in 2001. According to a DJV spokesman, the numbers have tripled in the last decade.

The raccoon has also come to populate cities in Germany. The first to receive the new city dwellers was Kassel, into which they started to move already fifty years ago. Kassel now has more than a thousand of them, corresponding to a density of more than 100 individuals per km^2 [8, 9]. Even if raccoon density is highest in Kassel, the raccoon has also established a significant presence in major cities such as Hamburg, Frankfurt and Berlin.

The raccoon population that developed east of Berlin, rooted at least partly in offspring of fur farm escapees from 1945 (as indicated by their darker-than-usual fur color), has spread into western Poland and established itself there [10].

Likewise, emerging raccoon populations in Denmark, the Netherlands, Belgium, Switzerland, Austria, the Czech Republic, and Slovakia are emigrants from Germany, as is, partly, the French population.

In some southern European countries, however, notably Spain, there are urban raccoon populations without obvious connections to the German ones [11, 12]. Genetic evidence indicates that at least two separate introductions had taken place in central Spain along the Jarama and Henares Rivers, each of them consisting of only a few individuals [13]. From those narrow bases, the populations have expanded and met, providing a larger genetic base for the rural Spanish raccoon population. In addition, it is assumed that the urban raccoons in, for example, Madrid, numbering several hundred and now subject to extermination attempts, represent other (pet) releases.

Besides those Central European releases, raccoons were also intentionally introduced to several parts of the then Soviet Union, notably Azerbaijan and Belarus in 1937 and 1954–1958, respectively, where stable populations were established [14]. So far, however, these populations are not connected to the Central European one and little is known about the numbers of raccoon found there. What is known, though, is that the sightings of raccoons there are increasing and sharply so.

Ecological Effects of Invasive Raccoons in Europe

There is an intensive ongoing debate in Europe, and particularly in Germany, about the ecological effects of the raccoon. On one side, in a unique confluence of interests, are hunters and (some part of) the environmental movement that see them as ecological vandals upsetting the ecosystem, reducing biodiversity, and threatening rare species with extinction. For example, in an interview, the leader of the Landesjagdverein said: "As nest-robbers, the raccoons are not only threatening pheasants and ducks, but for rare species like swamp turtles, they can be the difference between continued presence and disappearance" [15].

Capercailzie (wood grouse) and otter are also among the species seen as at risk from the raccoon [16], as are colonies of black-headed gulls and rare black storks [17, 18].

An often-repeated argument in favor of more intensive hunting is that the raccoon in Europe hardly has any natural enemies, since the large predators, such as wolves and lynx, are rare or absent. (The response of scientists such as Hohmann is that raccoons in North America also do not fall prey to natural enemies to any significant extent.)

However, there are some among hunters who hold a less-alarmist view of the raccoon, such as Claudia Gangl at the Bayerische Jagdverband [19].

Generally, as should be expected, scientists in the field take a more cautious view of the ecological effects of the raccoon in Europe. Pioneering studies in Germany by Lutz found that plant material made up 37% of raccoon food, invertebrates 40%, and vertebrates in the form of voles (*Microtus*), the rest [20]. He found no evidence of the raccoon having serious effects on the native fauna [21].

Studies of German raccoons in deciduous forests dominated by beech and oak showed that plant material dominated their diet, and that invertebrates constituted the bulk of the carnivore part [22].

Frank-Uwe Michler, who with a team of 20 biologists, veterinarians, and forest scientists studies the raccoons in Müntz National Park and has collected and analyzed 1500 sample of feces, said in a TV interview that "what the raccoon eats depends on what is easy available." In the National Park that means, depending on the time of the year, invertebrates such as earthworms, insect larvae, crustaceans, and water snails. Fruit, beechnuts, acorns, and berries also appear on the menu, as do small fish. "The raccoon is basically lazy, and eats what it can get with least effort. Newly-hatched chicks and eggs are not among those things - at least not in Müntz" [23].

However, in their study of raccoons in a wetland area in western Poland, Bartiszewicz and coworkers found a very different diet pattern. In their study, small rodents constituted the largest part (44%), followed by birds (15%), amphibians (13%), fish (13%), and insects (8%). Plant material was almost absent (2%), as were eggs (1%) [11].

The dire view of hunters as to the effects of raccoons on prey species finds support in some scientific studies, for example, those on the European swamp turtle (*Emys orbicularis*) in Brandenburg [24].

In a review of reports linking reproductive failures in bird colonies to time-correlated increases in raccoon abundance [25–27], Michler and Michler point out that such observations can be a basis for formulating a hypothesis to be scientifically tested, but do not in and of themselves suffice to conclude existence of a cause-effect relationship [28].

They also contrast single observations of nest-robbing raccoons, as well as experiments with arranged eggs in outplaced bird nests, with numerous examples of close proximity in time and place between raccoon resting sites and undisturbed nests of hollow-nesting bird species [29, 30].

They conclude their review by saying that so far there has been no scientific investigation that shows significant negative effect of raccoon predation on any native species and give reference to a long list of papers whose authors say the same [31].

Around the question of the effects on prey organisms of a predator that is also an omnivore, an interesting ecological debate has arisen concerning principles. On one side, the argument is made that an omnivore such as the raccoon will still be able to maintain its numbers when one of its prey species nears extinction, as it can then shift and live mostly from other food sources. Thus, it will be able to catch and kill even the very last of that particular species of prey. Specialized predators, on the other hand, drop in numbers when their prey becomes scarce, thereby reducing pressure and allowing the prey population to recover.

The argument on the other side is that an opportunistic feeder such as the raccoon, which is also lazy and will feed on whatever is easiest and most abundant, will thus balance an ecosystem rather than eliminate rare members of it.

If predation by raccoons has not, at least not yet, been shown to have any significant negative effect on the native prey species, could food competition have been a negative factor for other predators? Assessing that the wildcat (*Felis silvestris*) with its nocturnal hunting habits and climbing abilities would be the most direct food competitor to the raccoon and that the two species could also compete for sleeping and nesting sites, radiotelemetric studies of them were done in Solling, Niedersachsen. The results clearly showed that the two species had different room-time distributions and thus were not in hard competition [32].

Given its opportunistic nature, its ability to create niches for itself and its tactile-based hunting and food-search behavior, the raccoon is unlikely to be an overwhelming food competitor to badgers, foxes, or martens [22].

At a general level in Germany, areas in which the raccoon has been present in numbers for decades show no lower species diversities or densities than similar, raccoon-free areas [26, 33, 34].

The Raccoon as a Disease Transmitter in Europe

In one of the cases in which an emigrant species left some of its plagues behind when it emigrated, the European raccoon population has far fewer parasites than the population in North America. The most prevalent of the parasites that the raccoon did bring with it is the intestinal worm (*Baylisascaris procyonis*), which, as mentioned earlier, was prevalent in the original Hesse population but now can be found all over Europe, albeit with a low infection frequency rate, especially in the east of its range. There are cases in which this roundworm has been transmitted to humans, but they are rare, the symptoms relatively mild and the infection, when detected, was easily cured [35]. In all three cases, the raccoons involved were hand-fed pets [36].

As noted above with the raccoon in North America, the species is also both a carrier and a victim of rabies. However, the European version of the virus, the sylvatic one also called fox rabies, seems far less virulent to raccoons than the American one [37]. Since its introduction some eighty years ago, fewer than twenty cases of rabies-infected raccoons have been documented.

It has also been noted that raccoons in captivity infected with European rabies do not show the typical aggressive behavior, but withdraw to their hiding places [38].

One disease for which the raccoon can become epidemiologically important and that could affect the future population of the species itself is distemper. In North America, distemper is an important factor controlling local raccoon populations and, of and on, reducing them substantially. In Germany, the first eight cases of distemper among wild raccoons were reported in 2007 [39]. So far, however, only females have been documented to have succumbed to the disease. Males have been found living well, with lots of antibodies as proof of infections they have survived.

There are few studies in Europe of the causes of death among raccoons in the wild. For one such study, in Müritz-Nationalpark, 68 animals (23 adult females, 28 adult males, and 17 youths) were fitted with UKW necklace senders for radiotelemetric surveillance and followed from 2006 to 2009. Of these, 22 died during the period of observation (32%). Three were roadkills on a highway through the park, and two were shot after having migrated out. Five died of distemper and two of brain parasites. Of the remaining ten, six died of different accidents, and for four, the cause of death could not be determined [40].

Obviously, both roadkill and hunting can be expected to be larger causes of death among raccoons outside of national parks and other protected areas, but the relatively high proportion of virus and parasite deaths are anyhow noteworthy.

Economic Aspects of Raccoons in Europe

Raccoons love sweet berries and fruit such as cherries, plums, and grapes, so when such delicacies are available they may gather in farther larger groups than they do normally, to enjoy the offering. As this also occurs in the autumn, at a time when they need to fatten themselves up in preparation for the winter hibernation, they become voracious consumers.

Obviously, they then incur an economic cost to fruit plantations. The total extent of the damage has not been assessed, but its significance is seen as only local [28]. Should the population density of the raccoon living in wine-growing districts continue to increase, the damage would in all likelihood become of concern.

More significant, then, as of now, is the damage that city-dwelling raccoons cause, not only to suburban fruit tree gardens, but first and foremost to buildings. Their joint latrines in the attics of villas and apartment buildings can give rise to significant cleanup and restoration costs. Furthermore, their ransacking of garbage cans, interference with clothes hung out to dry, and demolition of compost heaps do not only cause irritation but also have an economic side. In Kassel, where the density of raccoons is especially high, a number of measures have been taken to reduce the disturbances with some degree of success [9], but not without costs.

Ausländisch oder heimish? (Alien or Native?)

In Germany, the raccoon is officially "heimisch" (native or indigenous). The status is recognized in federal law, Bundesrecht, and follows from what is stated in BNatSchG § 7 Abs. 2 Nr. 7. Interpreted liberally, the rule says that an animal (or plant) that in nature and without human support can survive over several generations is to be regarded as "heimisch" even when it was first introduced by humans.

As both heimisch and wild, the raccoon falls under Jagdrecht and can be hunted in all Bundesländer except Bremen and Saarland.

In Poland, too, the raccoon is covered in the hunting legislation since 2004 and can be shot or trapped from July 1 to March 31.

The Future of Raccoons in Europe

There is little doubt that the rapid expansion of the raccoon population in Europe will continue, and that new areas and countries will be added to its range of distribution. The only possible hindrance looming on the horizon is if distemper secures a permanent hold and keeps the population density down with periodic outbreaks of mass mortality. That scenario aside, though, the raccoon seems still to be only in the early phase of its conquest of Europe.

So far, if the early population counts and estimates are takes *ad notam*, the annual raccoon population growth in Europe since introduction would on an average have been 12–17%, meaning a doubling every 3–5 years. Should that continue to be the case, the population would pass two million within a decade and, maybe, ten million by 2040. However, one would expect the growth rate to slow down, even if there are few signs of that yet.

References

1. Hohmann, U., Bartussek, I., & Böer, B. (2001). *Der Waschbär* (in German). Reutlingen, Germany: Oertel+Spörer.
2. Lutz, W. (1996). The introduced raccoon Procyon lotor population in Germany. *Wildlife Biology, 2*, 228.
3. Holmgren, V. C. (1990). *Raccoons: In folklore, history & today's backyards* (p. 157). Santa Barbara, CA: Capra Press.
4. Davis, H. B. (1907, October). The Raccoon: A study in animal intelligence. *The American Journal of Psychology* (Champaign, Illinois: University of Illinois Press), *18*(4), 447–489.
5. Zeveloff, S. I. (2002). *Raccoons: A natural history*. Washington, DC: Smithsonian Books.
6. Pomeroy, L. W., Bjørnstad, O. N., & Holmes, E. C. (2008). The evolutionary and epidemiological dynamics of the paramyxoviridae. *Journal of Molecular Evolution, **66***(2), 98–106. Retrieved from http://en.wikipedia.org/wiki/Canine_distemper - cite_ref-Pomeroy2008_5-1.
7. From Society for Conservation Biology. More Raccoons may mean fewer songbirds, 28 July 2003, *ScienceDaily*.
8. Frank-Uwe Michler: Untersuchungen zur Raumnutzung des Waschbären (Procyon lotor, L. 1758) im urbanen Lebensraum am Beispiel der Stadt Kassel (Nordhessen). 25. Juni 2003, S. 108 (online (PDF; 3,9 MB). Accessed 10 April 2015.
9. Michler, F.-U. (2004). Waschbären im Stadtgebiet. – Wildbiologie International 5/12, Infodienst Wildbiologie & Oekologie. – Zürich, Schweiz, 16 S.
10. Bartoszewicz, M., Okarma, H., Zalewski, A., & Szczesna, J. (2008). Ecology of the raccoon (Procyon lotor) from western Poland. *Annales Zoologici Fennici, 45*, 291–298.
11. Beltrán-Beck, B., García, F. J., & Gortázar, C. (2012). Raccoons in Europe: Disease hazards due to the establishment of an invasive species. *European Journal of Wildlife Research, 58*, 5–15.
12. García, J. T., García, F. J., Alda, F., González, J. L., Aramburu, M. J., Cortés, Y., Prieto, B., Pliego, B., Pérez, M., Herrera, J., & García-Román, L. (2012). Recent invasion and status of the raccoon (Procyon lotor) in Spain. *Biological Invasions*. doi:10.1007/s10530-011-0157-x.
13. Alda, F., Ruiz-López, M. J., García, F. J., Gompper, M. E., Eggert, L. S., & García, J. T. (2013). Genetic evidence for multiple introduction events of raccoons (*Procyon lotor*) in Spain. *Biological Invasions, 15*(3), 687–698.

14. Heptner, V. G., & Sludskii, A. A. (2002). *Mammals of the Soviet Union. Vol. II, part 1b, Carnivores(Mustelidae & Procyonidae)*. Washington, DC: Smithsonian Institution Libraries and National Science Foundation. ISBN:90-04-08876-8.
15. Potsdamer neueste Nachrichten 13/5 2013.
16. Eher unerwünscht: Waschbären in Mecklenburg-Vorpommern. www.erlebe-mv.de/natur/waschbaer.htm
17. Waschbären haben keine Feinde. Jäger Hubertus Ostermann kämpft gegen die Ausbreitung der Allesfresser Volkstimme.de 21.06.2013 02:12 Uhr.
18. DJV. (2012). Tierische Invasoren breiten sich schnell aus. – Pressemitteilung des Deutschen Jagdschutzverbandes vom 8.06.2012.
19. Retrieved from http://www.jagd-bayern.de/fileadmin/BJV/Jagd_In_Bayern/jib_2014_02_14_waschbaer_biologie.pdf
20. Lutz, W. (1980). Teilergebnisse der Nahrungsanalysen am Waschbären (Procon lotor), Nordhessen. *Zeitschrift für Jagdwissenschaft, 26*, 61–66.
21. Lutz, W. (1981). *Untersuchungen zur Nahrungsbiologie des Waschbären (Linné 1758) und zum Einfluß auf andere Wildarten in seinem Lebensraum*. Dissertation Universität Heidelberg.
22. Hohmann, U., Gerhard, R., & Kasper, M. (2000). Home range size of adult raccoons (Procyon lotor) in Germany. *Zeitschrift für Säugetierkunde, 65*, 124–127.
23. Expeditionen ins Tierreich. Waschbären – Einwanderer aus Wild West. Mittwoch, 30.10.2013, **NDR**, 20:15 Uhr.
24. Schneeweiss, N., & Wolf, M. (2009). Neozoen – eine neue Gefahr für die Reliktpopulationen der Europäischen Sumpfschildkröte in Nordostdeutschland. *Zeitschrift für Feldherpetologie, 16*, 163–182.
25. Fischer, S., & Dornbusch, G. (2005). Bestandssituation ausgewählter Brutvogelarten in Sachsen-Anhalt – Jahresbericht 2004. – Berichte des Landesamtes für Umweltschutz Sachsen-Anhalt. *Sonderheft, 1*, 3–23.
26. Schneider, R. (2012). Achtung, Panzerknacker. Waschbär – Bedrohung für heimische Arten. *Unsere Jagd,* (6), 30–33.
27. Helbig, D. (2011). Untersuchungen zum Waschbären (Procyon lotor Linné, 1758) im Raum Bernburg. *Naturschutz im Land Sachsen-Anhalt, 48*(1 + 2), 3–19.
28. Fritz Michler, F.-U., & Michler, B. A. Ökologische, ökonomische und epidemiologische Bedeutung des Waschbären (Procyon lotor) in Deutschland – eine aktuelle Übersicht. *Beiträge zur Jagd- und Wildforschung, 37*(2012), 387–395.
29. Horstmann, D., & Schmincke, B. (2004). Beobachtungen zur Koexistenz von Mittelspecht und Waschbär. *Natur und Heimat, 64*(2), 55–60.
30. Hohmann, U., & Bartussek, I. (2011). *Der Waschbär* (3. aktual. Aufl., 200 S). Reutlingen: Oertel u. Spörer.
31. Becker, S. (2011). Der Waschbär (Procyon lotor Linné, 1758) in Deutschland – eine Konfliktart? Bachelorarbeit Georg-August-Universität Göttingen, 70 S.
32. Hohmann, U., & Hupe, K. (1998). Interspecific competition of the raccoon (Procyon lotor) and the wildcat (Felis silvestris silvestris) with regard to rest sites in Germany. In: C. Thomaidis & N. Kypridemos (Eds.) *Agriculture forestry – game, integrating wildlife in land management, IUGB-Proceedings Thessaloniki* (pp. 361–367).
33. Hohmann, U. (2001). Stand und Perspektiven der Erforschung des Waschbären in Deutschland. *Beiträge zur Jagd- und Wildforschung, 26*, 181–186.
34. Lutz, W. (1995). Erfahrungen mit ausgewählten Säugetierarten und ihr zukünftiger Status. In: *Akademie für Natur- und Umweltschutz Baden-Württemberg* (Hrsg.): Statuskolloquium "Neozoen – neue Tierarten in der Natur". Tagungsführer. – Fellach (pp. 297–310).
35. Michler, F.-U., & Köhnemann, B. A. (2008, Mai) Ökologische und ökonomische Bedeutung des Waschbären in Mitteleuropa – Eine Stellungsnahme
36. Bauer, C. (2011). Baylisascariose (Baylisascaris procyonis) – eine seltene parasitäre Zoonose in Europa. *Berliner und Münchener Tierärztliche Wochenschrift, 124*(11/12), 465–472.

37. Vos, A., Ortmann, S., Kretzschmar, A. S., Köhnemann, B., & Michler, F.-U. (2012). The raccoon (Procyon lotor) as potential rabies reservoir species in Germany: A risk assessment. *Berliner und Münchener Tierärztliche Wochenschrift, 125*(5/6), 228–235.
38. Dröscher, V. (1979). Waschbären erobern Deutschland. *Das Tier, 19*(2), 20–23.
39. Michler, F.-U., Köhnemann, B. A. (2008, June). Erste Ergebnisse. *Projekt Waschbär.*
40. Michler, F.-U., Köhnemann, B. A., Roth, M., Speck, S., Fickel, J., & Wibbelt, G. (2009). Todesursachen sendermarkierter Waschbären (Procyon lotor L.,1758) im Müritz-Nationalpark (Mecklenburg-Vorpommern). *Beiträge zur Jagd- und Wildforschung, 34*, 1–17.

Canadian Beavers in Tierra del Fuego

Beaver with Sour Cream Ingredients:

2–4 lb beaver (without bones)
1/2 cup flour
1 tsp. salt
1/4 tsp. paprika
1/2 tsp. salt
1/2 cup water
1 cup sour cream
Oil for frying
1 onion

Directions:

Clean beaver and soak overnight in salted water (1 tbsp. salt to 1 quart water). Drain, cut into cubes, and roll in 1/2 cup flour seasoned with 1 tsp. salt and 1/4 tsp. paprika.

In a large skillet, fry beaver in fat until browned on all sides, and then cover with sliced onion. Sprinkle the onion slices with 1/2 tsp. salt. Add 1/2 cup water. Cover the skillet tightly and simmer for 1 h.

Add 1 cup sour cream the last 15 min of cooking time.

Recipe serves 2–4, depending on the size of the animal.

Country-Style Beaver Ingredients:

2–3 lb beaver steaks, 1/2 in. thick
Bacon fat
2 cups flour
1 tsp salt

A. Jernelöv, *The Long-Term Fate of Invasive Species*,
DOI 10.1007/978-3-319-55396-2_16

1 tsp pepper
2 medium onions
1 can cream-of-mushroom soup
1 can or 1/2 lb fresh-sliced mushrooms

Directions:

Combine flour, salt, and pepper in a sealable bag or 2 quart sealable plastic container, and shake until mixed. Add beaver and shake until well coated. Save remaining flour mixture. Dice onions. Melt enough bacon fat in the bottom of a frying pan to sauté onions and beaver. Sauté onions and floured beaver in bacon fat, adding more fat as needed. After sautéing, place beaver aside.

Combine soup and mushrooms in frying pan. Dissolve 2–3 heaping tbsp. of seasoned flour in 2 cups cold water. Add to soup mix and simmer for 5 min. Add beaver and onions to mix and simmer covered for 30 min.

In the late 1990s, the governments of Argentina and Chile, frequently at odds politically, nevertheless, launched a joint attempt to control the beaver population on Tierra del Fuego, the archipelago that the two countries share, at their southernmost tips. This involved not merely paying hunters to kill the animals but also encouraging restaurants to serve beaver meat in gourmet dishes based on recipes such as the ones above [1].

Introduction and Spread of the Canadian Beaver

A 1946 newsreel, which in those days played in movie theaters across Argentina before the start of the main feature, shows the release of beavers, trapped in Canada and flown to Tierra del Fuego, into the region's largest lake, Lago Fagnano. While the camera pans vast native forests, the narrator explains that these valuable fur-bearing animals "eat bark and branches of trees, which they will find in abundance here." When the animals dive down and disappear out of sight in their new environs, the narrator concludes that "they are now in the hands of God" [2]. And God, as it turned out, was good to them.

The background to the release was that the new Argentine government of Juan Perón wanted to foster a fur trade in the economically lackluster territory. Tom Lamb, the pilot who organized the catch in Canada and flew the beavers to Argentina, was remunerated with US$650 per animal [3]. In the newsreel referred to above, the number of imported beavers was given as 20. Other sources mention 50 (or 25 breeding pairs) [4].

Release of Manitoba beavers into Tierra del Fuego, 1946 (picture from LambAir.com)

The subsequent spread of the beavers was not particularly rapid, but it was steady. Significant ecological and landscape changes were noted as the beavers advanced, but it wasn't until they crossed over to the South American mainland that the invasion threat was taken seriously. That was when the bounties and beaver recipes mentioned above came about. Most control efforts were centered along the edges of the mainland to slow the spread there. Argentina had a goal of trapping 7000–10,000 beavers per year, a number Chile exceeded with 11,700 killed between July 2005 and June 2006. In Tierra del Fuego and the many uninhabited islands in its archipelago, however, not much hunting took place despite the incentives. By the time the bounty payments stopped, the Argentine goals had not been reached and the Chilean catches had not been sustained [2].

Despite these efforts, the beaver population continued to grow. Whereas in 2000, the number of beavers was estimated to be 100,000 [5]; by 2008 it was 200,000 [6] and in 2010, 250,000 [3].

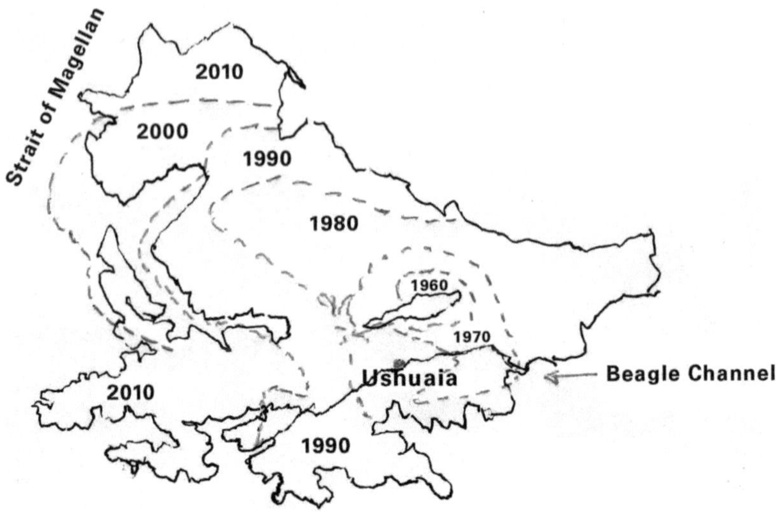

Spread of beavers over Tierra del Fuego

Practically all articles on the subject stress the absence of predators on Tierra del Fuego as a factor behind the sharp increase in beaver numbers over the nearly seven decades since their introduction.

The Canadian Beaver (*Castor canadensis*)

The Canadian beaver is a large, semiaquatic rodent. It weighs between 10 and 30 kg as an adult, with 20 kg as the norm. The length of the head and body is 75–90 cm, with the flat tail adding another 20–35 cm. Old beavers can exceptionally be significantly larger with lengths well over a meter and weights up toward 50 kg.

The beaver has many traits adapted to a life in water, such as a flat, paddle-shaped tail; webbed, fin-like hind feet; membrane-covered eyes that allow underwater vision; and nostrils and ears that are sealed when submerged. A thick layer of subcutaneous fat and fur with long, coarse outer hairs and short fine inner hairs, made waterproof by an oily substance known as castoreum, insulate the body from wetness and cold. This "beaver oil" is secreted from scent glands near the genitals and is, together with its fur, a valuable commodity for which the animal is hunted.

A descendant of the original beavers transplanted from North America to Patagonia in 1946 chews on a lenga branch (photo credit to J. Ducnuigeen [2])

The beaver has strong incisor teeth and relatively small, highly mobile front paws with claws. They are used to fell trees by gnawing in a characteristic way and to manipulate the resulting logs to form the dams and lodges that are the trademarks of the beaver.

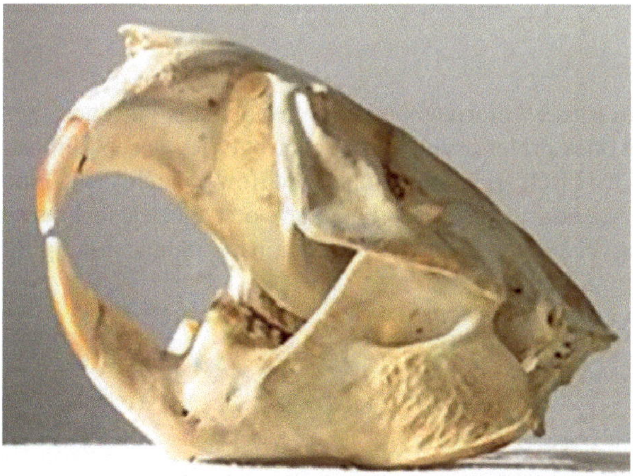

Skull of a North American Beaver found on San Francisco Bay shore (from Wikipedia)

Beaver gnaw on a fallen birch tree (from http://www.alltomtradgard.se/profil/album/bildsida/?albumId=156011&imgId=175884)

Beavers are true vegetarians and eat leaves, buds, and the inner bark of trees. The Canadian ones prefer aspen and poplar, but they accept almost any hardwood species such as birch, red oak, maple, willow, beech, ash, black cherry, and hornbeam. If necessary, they'll also eat softwood species such as spruce and pine. In early spring, beavers will also feast on aquatic plants, among them water lilies and cattails.

Beaver dam, northern California, USA (photo from Wikipedia)

Beaver lodge, Ontario, Canada (photo from Wikipedia)

When suitable trunk trees are not available, beavers may use rocks and stones to build their dams.

A beaver dam made of stones. Bear Creek, Alpine Meadows, California, USA (photo from Wikipedia)

Beavers build dams to secure an underwater entrance to their lodges and to provide easy and safe access in water to feeding grounds and logging sites. The dams need to be deep enough not to bottom freeze in winter.

Canadian beavers pair up for life. They typically have a litter with two or three kits a year, although up to six kits have been recorded. Normally, they are sexually mature at the age of 3 years and live with their parents till then, forming family units of up to a dozen members. They are highly territorial.

A close relative of the Canadian beaver (*Castor canadensis*) is the European beaver (*Castor fiber*). However, the species have different numbers of chromosomes, 40 and 48, respectively, and do not interbreed.

In North America, bears and wolves are regular predators of beavers, especially the young ones, which are also occasionally taken by coyotes, foxes, and large birds of prey such as eagles. Predation, however, is not seen as a major cause of death for beavers.

Ecological Effects of Beavers in Its Native Range

The most important ecological results of the presence of the Canadian beaver are the aquatic ecosystems and wetlands formed by their damming of waterways. A number of studies have found that beaver ponds are favorable habitats for young salmonids. In Sweden, brown trout growing up in beaver ponds were larger than those from river sections and small streams. Larger trout survived periods of drought by retreating to them [7].

Similarly, an American study of brook trout, rainbow trout, and brown trout in Sagehen Creek in the northern Sierra Nevada at an altitude of 1800 m found that beaver ponds increased the number of fish, their size, or both [8].

Studies in the Stillaguamish River in Washington State found that extensive loss of beaver ponds resulted in an 89% reduction in coho salmon smolt production [9].

It's sometimes claimed that beaver dams might hinder salmon migration. In a meta-review of such claims, it was found that 78% of them were not supported by data [10]. On the contrary, a number of studies found no such effects. Radiotelemetry-tagged cut trout and brook trout in a Utah study were found to cross beaver dams in both directions, including up to 2 m-high dams [11]. In Alaska, coho salmons jumped beaver dams of similar heights and were found above all of them. They were also found in higher densities in streams with beavers than in those without them [12]. Downstream migration of Atlantic salmon smolts was unaffected by beaver dams even in periods of low water flows [13].

In the presence of beavers, the areas of open waters and wetlands in the landscape increase significantly. In an Alberta study, using 12 aerial photo mosaics from 1948 to 2002, the impact of the return of beavers on open water areas found a ninefold increase [14].

Beaver ponds store runoff water from snowmelt and rainy seasons and distribute it over the year, thereby increasing stream flows in dry seasons. Percolation from beaver ponds also raises groundwater tables. In the Alberta study, the authors concluded that beavers have a dramatic influence on the creation and maintenance of wetlands even during extreme drought [15].

By providing settling basins, beaver ponds also improve stream water quality by removing suspended matter, nutrients, and pollutants as shown by studies from the Maryland coastal plain and the Lake Tahoe area in California [16, 17]. In addition, fecal coliform and streptococci bacteria excreted into streams by grazing cattle are reduced by beaver ponds, where slowing currents lead to settling of the bacteria in bottom sediments [18].

The increased areas of open waters that beavers create also help waterfowl. In a study in Wyoming, it was found that watercourses with beavers had 75 times more ducks than those without [19]. Some species, like trumpeter swans and Canada geese, often prefer beaver lodges as nesting sites [20].

There are also several other ways in which beavers benefit birds that frequent their ponds. By removing trees at the pond's sides, they increase the density and height of grasses and shrubs that improve waterfowl nest cover. Coppicing of waterside trees that respond with intensive shoot production, such as willows and cottonwoods, has similar effects.

Forest gaps, where beavers have fallen the trees, and the "gradual edge," the transition zone from pond to forest with grasses, forbs, and shrubs of increasing average height, are attractive to migrating birds. Dead trees, drowned in the beaver impoundments, become feeding and nesting sites for woodpeckers. The cavities they make attract many other bird species such as flycatchers, tree swallows, tits, and wood ducks [21]. Fish-eating birds such as herons, egrets, cormorants, mergansers, and kingfishers utilize the beaver ponds for fishing [22].

Through its logging, the Canadian beaver will have an effect on the size of trees on the waterfronts and some tens of meters "inland." Some of the effects are discussed above, but also the composition and diversity of tree species can be affected. A study of ten streams in the Lake Tahoe basin using multispectral video graphics showed that deciduous, thick, and thin herbaceous vegetation was more highly concentrated near beaver dams, whereas conifer tree abundance decreased [23].

Bridge Creek, a stream in semiarid eastern Oregon, has been the object of a couple of long-term studies of beaver dams and riparian vegetation. The conclusion is that within a few years after arrival of beavers, "dense stands of woody plants of greater diversity occupied a larger portion of the floodplain" and that, therefore, "beaver appear to increase riparian vegetation, given enough years to aggrade sediments and pond heights sufficiently to create widened, well-watered riparian zones, especially in areas of low summer rainfall" [24, 25].

In North America, the Canadian beaver was ubiquitous before the fur trade drove it close to extinction. They inhabited the continent from the Atlantic to the Pacific and from the Arctic tundra to the deserts of northern Mexico. The estimates of the population size in those days range from 60 million [26] to 100–200 million [27]. The many ponds they created have been described as "moth holes" in the dense forests, and the rich fish and bird life in and around them, in addition to the beavers themselves, attracted Indian hunters as well as trappers. The bottoms of old, abandoned, or emptied beaver ponds also found some use in small-scale agriculture, as the sediment soil there was more fertile.

When intensive hunting had reduced the population of *Castor canadensis* to small isolated subpopulations, mostly in remote areas and counting only a million or so animals in total, the species became partly protected in the late nineteenth and early twentieth century. At the same time, the price of fur and castoreum dropped to levels that made hunting unprofitable. Beaver numbers have since rebounded, and it is estimated that the current North American population is 10–15 million [27].

The overall conclusion from studies of the ecological effects of the beaver in its native habitat is that it is a keystone species that maintains the aquatic landscape with its water surfaces and wetland areas. When it was hunted to rarity, the landscape became more homogenous and impoverished, and, consequently, species diversity and richness decreased. Thus, reintroduction of the beaver from areas where it is still missing is seen as environmentally desirable.

Nevertheless, economically, the resurgence of the beaver in North America has resulted in some damage. Already, the larger trees that beavers can gnaw through represent a value, but far more important are the effects of flooding caused by the dams. Collectively, they cause death or retarded growth of trees over large areas, as well as destruction of roads and railroads. Beaver burrowing damages levees and man-made dams. Damage caused by beavers in 16 southern US states was estimated to be $4 billion over 40 years [28]. Assuming an average population of two million beavers in those states during those years, that would amount to an average economic damage per beaver of $50 a year. Other assessments are lower. An often-quoted sentence in connection with beaver control programs is: "Some southeastern states where beaver damage is extensive have estimated the cost at $3 million to $5 million dollars annually for timber loss; crop losses; roads, dwellings, and flooded property; and other damage". For example, [29] I've been unable to find the original article, but if this often-referred-to figure is used, it would indicate an annual damage cost per beaver of $5 or less.

Beavers as Symbols and Mascots

Castor canadensis is the national animal of Canada and one of its national symbols. As such, it was depicted on the first Canadian stamp, the "three-penny beaver," in 1851, as well as on the current 5 cent coin.

The first **Canadian** postage stamp was issued on April 23, 1851. It was designed by Sir Sandford Fleming and figured the industrious beaver

In addition, the beaver is the state animal of New York and Oregon and is on the coat of arms of the Hudson Bay Company, University of Toronto, and London University, as well as of New York City and its state capitol, Albany. Beaver images are also very commonly found on the emblems of engineering schools, including such leading ones as MIT (Massachusetts Institute of Technology) and CALTECH (California Institute of Technology). It figures as a mascot for Oregon State University, City College of New York, and Babson College. Also for the Montreal Summer Olympics 1976, the beaver was the mascot under the nickname amik (which means beaver in the Anishinaabe language).

Tierra del Fuego

Tierra del Fuego is the southernmost tip of South America, south of the Straits of Magellan. It is an archipelago with one large island, Isla Grande, with an area of 48,100 km^2, and a large number of small ones, including Cape Horn. Tierra del Fuego is divided between Chile and Argentina, along the meridian 68°34′ west from the Straits of Magellan to the Beagle Channel, with Argentina's lands on the east side.

The region has a subpolar oceanic climate with short, cool summers (3 months, 9 °C) and long, mild winters (6 months, 0 °C). On the western side, precipitation is high, up to 3000 mm annually, but decreases quickly as one moves eastward. In the northeast there are steppes and cool semi-deserts. About a third of the area is forested.

There are only six species of native trees on Tierra del Fuego. One of them is a conifer of the cypress family (*Pilgerodendron uviferum*). Another, called "winter's bark," is thought to have medical qualities and was used by sailors against scurvy. Three tree species are southern beeches of the genus *Nothofagus*. One of them, called "lenga," is often the dominating forest-forming one.

Of native flowering plants, some of the most well-known ones are the black bush of Tierra del Fuego, with small daisy-like flowers; bell biflora with white and purple bell-shaped flowers; and the strawberry of Magellan with sweet, blackberry-like fruits [30].

Another common plant is the Tierra del Fuego dock (*Rumex fueginus*).

Generally speaking, however, there are few native plant species. That is also the case for non-marine vertebrates. One of them is the guanaco (*Lama guanicoe*), a wild relative of the llama. Elsewhere in its native range in the southern Andes, it has been hunted to relative rarity. That was also the case in Tierra del Fuego, and in the 1970s only a few thousand remained on Isla Grande. There has since been a remarkable recovery, and the Chilean population on Tierra del Fuego is now estimated to be over 150,000. Add to that whatever the Argentine population might amount to (50,000?), and it is clear that this may be a historic high number as landowners become concerned about the potential damage to forestry and sheep-grazing areas [31].

Guanacos crossing the road in Tierra del Fuego, Chile. Government-sanctioned culling has been opposed by animal rights activists and tourism officials (Credit Tomas Munita) [31]

In fact, alien mammalian species outnumber native ones 18:11. There are only four native species of freshwater fish, according to a large-scale overview study [32], of which only one, "El puye" (*Galaxias maculatus*), is regularly reported from surveys. There is an equal number of introduced salmonids (4), but the latter dominate in number and size. Of these, the brown trout (*Salmo trutta*) has established a sea-migrating population in Rio Grande that is world famous among fly fishermen.

Ecological Effects of Beavers on Tierra del Fuego

Also in their new habitat, the beavers gnawed trees, built dams, and flooded areas, creating new water surfaces and wetlands. Somehow, though, the destruction seemed more widespread and intensive and the positive effects less noticeable than in their native home in North America [33].

Beaver dam in Tierra del Fuego (photo from Wikipedia Ilya Haykinson)

Dead trees as a result of the construction of a beaver dam (photo from Wikipedia)

There are several reasons, regularly repeated in articles on the subject, that help explain why the forests in Tierra del Fuego are more prone to extensive beaver damage than the North American ones: The tree species native to Tierra del Fuego do not produce new shoots from the root when the stem is gnawed off, unlike aspen and willows, for example, and their roots are much less tolerant to inundation. According to scientists at CADIC (Centro Austral de Investigaciones Cientificas), 50% of Tierra del Fuego's riparian forests are now damaged by beavers [2, 34].

Christofer Anderson, an ecology professor at the Universidad de Magallanes and an author of several studies of the beaver on Tierra del Fuego, claims that "the changes in the forested portion of this biome is the largest landscape-level alteration in the Holocene – that is for some 10,000 years" [35].

Beaver-engineered landscapes in Tierra del Fuego (pictures from [36])

In an extensive review of studies published both in the open and gray literature in Chile and Argentina about the impact of invasive species on native ones and on ecosystems in Tierra del Fuego [37], 58 out of 85 publications (68%) dealt with the beaver.

These studies provided an ample body of evidence showing ecosystem impacts, including effects on nutrient cycling and soil properties [38] and stream food webs [39–41].

Beavers also facilitate other invasive species such as exotic plants in meadows [42].

Noteworthy is that these are the same effects of beaver ponds on erosion, sediment and nutrient transport, sedimentation, soil fertility, etc. as those seen in North America. There, ecologists see them as positive. In Tierra del Fuego, however, they see them as negative. The reason for the dichotomy is, of course, that in the view of orthodox ecologists, the world was perfect without humans and that man's impact, by definition, is always negative.

The question of beaver ponds and native fish gets a little complicated by the presence of the introduced brown trout. El puye (*Galaxias maculatus*) benefits from beaver ponds, but so do brown trout and the other alien salmonides, in Tierra del Fuego as well as in North America and Scandinavia. They in turn reduce the abundance of el puye, especially in beaver ponds [40]. However, in summary, also considering the predation by trout and salmon, beaver damming favors the native fish [36]. Other studies found biodiversity in freshwater systems unaffected even when beavers caused major landscape alterations [43].

A question, that has not been addressed in any study that I have found, is whether the rich pastures of often alien plants on the bottoms and sides of former beaver ponds have anything to do with the remarkable recovery of the native wild llama, the guanaco.

The Beaver as a Tourist Attraction

While ecologists and forest landowners on Tierra del Fuego see the beaver as a pest, the tourist industry has an entirely different view of the animal. They watch the increasing numbers of mainly upmarket tourists passing Ushuaia on their way to and from package tours to Antarctica, spending a night in transit at some hotel in town, and calculate that if the visitors could be convinced to stay an extra day or two, much tourism revenue could be earned. This is where the beaver comes in. The tourists generally either have a positive view of beavers or know nothing at all about them. Either way, they might like to see them. Thus, the beaver has been promoted as a symbol for the area, and many shops in Ushuaia have a beaver in their signs. Beaver mascots compete with penguin ones, both as stuffed animals for kids and as key-ring appendices. Beaver paintings can be found in practically all hotels, most restaurants, and several offices of local authorities. The manufacture of beaver hats and gloves is a small but growing local business. The local ski resort is called Beaver Mountain. A popular excursion into the Tierra del Fuego National Park features beaver viewing, and photo safaris focused on beavers are as popular as those for penguins. The organizers of the world-renowned brown trout fishing tours are well aware of the role that beaver ponds play for the brown trout smolt production in Rio Grande and its tributaries. To them the animals are an asset and contribute to the economic development, not exactly the way it was meant to be through sales of fur and castoreum, as president Juan Peron thought, when he authorized their introduction, but still.

Beaver symbols as tourist traps (pictures from [4])

A Joint Argentine-Chilean Eradication Attempt

The collaboration between the two reluctant partners, the governments of Argentina and Chile, which begun in the late twentieth and early twenty-first centuries for the purpose of controlling the beaver population at Tierra del Fuego and preventing the spread of the invader to the South American mainland, took a further step forward in 2008, when they agreed, in principle, to launch a full eradication program. A number of pre-studies of the economics of the undertaking and possible methods were carried out. A feasibility study [44] recommended a 9-year program with professional trappers, specialized dogs, and helicopter gunships with beaver hut-breaking bombs and boats, all of which would attack the beavers on a rolling front, aided by satellite images. The preliminary price tag was set at US$35 million. In a *Nature* article, it was called "the largest eradication program ever attempted" [34].

Not everyone is convinced of the feasibility of the program, given the rugged nature, the many uninhabited islands, and the inhospitable climate of Tierra del Fuego. The ever-more economically important tourist industry also has major misgivings about the desirability of the venture.

So far, the full program has not been launched.

References

1. Retrieved February 9, 2014, from. http://www.cajuncookingrecipes.com/wildgamerecipes/
2. Worth, K. (2014, March 10). Argentina and Chile decide not to leave it to beavers. *Scientific American.*
3. Retrieved February 9, 2015, from http://thingstodo.viator.com/argentina/beaver-problems-in-tierra-del-fuego/
4. Retrieved February 9, 2015, from https://jonsadventure.wordpress.com/2013/02/07/history-lesson-1-beaver-invasion-of-tierra-del-fuego/
5. Mok, K. (2008). Plans to totally eradicate Tierra del Fuego's invasive beavers. Natural Sciences. Retrieved February 9, 2015, from http://www.treehugger.com/natural-sciences/plans-to-totally-eradicate-tierra-del-fuegos-invasive-beavers.html
6. Forero, J. (2011). Imported beavers gnaw away at Argentina's forests. Retrieved February 9, 2015, from http://www.npr.org/2011/06/24/137067240/imported-beavers-gnaw-away-at-argentinas-forests
7. Hägglund, Å., & Sjöberg, G. (1999). Effects of beaver dams on the fish fauna of forest streams. *Forest Ecology and Management, 115*(2–3), 259–266.
8. Gard, R. (1961). Effects of beaver on trout in Sagehen Creek, California. *Journal of Wildlife Management, 25*(3), 221–242.
9. Pollock, M. M., Pess, G. R., & Beechie, T. J. (2004). The importance of beaver ponds to coho salmon production in the Stillaguamish River Basin, Washington, USA. *North American Journal of Fisheries Management, 24*(3), 749–760.
10. Kemp, P. S., Worthington, T. A., Langford, T. E. L., Tree, A. R. J., & Gaywood, M. J. (2011, June). Qualitative and quantitative effects of reintroduced beavers on stream fish. *Fish and Fisheries, 13*(2), 158.
11. Lokteff, R. L., Roper, B. B., & Wheaton, J. M. (2013). Do beaver dams impede the movement of trout? *Transactions of the American Fisheries Society, 142*, 1114–1125.

12. Bryant, M. D. (1984). The role of beaver dams as coho salmon habitat in southeast Alaska streams. In J. M. Walton & D. B. Houston (Eds.), *Proceeding, Olympic Wild Fish Conferences* (pp. 183–192). Port Angeles, WA: Peninsula College, Fisheries Technology Program.
13. Collen, P., & Gibson, R. J. (2001). The general ecology of beavers (*Castor* spp.) as related to their influence on stream ecosystem and riparian habitats, and the subsequent effects on fish—A review. *Reviews in Fish Biology and Fisheries, 10*, 439–461.
14. Hood, G. A., & Bayley, S. E. (2008). Beaver (*Castor canadensis*) mitigate the effects of climate on the area of open water in boreal wetlands in western Canada. *Biological Conservation, 141*, 556–567.
15. University of Alberta. Busy beavers can help ease drought, 26 February 2008. *ScienceDaily*. Retrieved from www.sciencedaily.com/releases/2008/02/080220130511.htm
16. Correll, D. L., Jordan, T. E., & Weller, D. E. (2000, June). Beaver pond biogeochemical effects in the Maryland Coastal Plain. *Biogeochemistry, 49*(3), 217–239.
17. Muskopf, S. (2007). *The effect of beaver (Castor canadensis) dam removal on total phosphorus concentrations in Taylor Creek and Wetland, South Lake Tahoe, California*. Thesis, Humboldt State University, Natural Resources. Retrieved February 10, 2015.
18. Skinner, Q. D., Speck, J. E., Smith, M., & Adams, J. C. (1984, March). Stream water quality as influenced by beaver within grazing systems in Wyoming. *Journal of Range Management, 37*(2), 142–146.
19. McKinstry, M. C., Caffrey, P., & Anderson, S. H. (2001). The importance of beaver to waterfowl and wetlands habitats in Wyoming. *Journal of the American Water Resources Association, 37*(6), 1571.
20. Brenner, F. J. (1960). Canada geese nesting on beaver lodge. *The Auk, 77*(4), 476–477.
21. Rosell, F., Bozser, O., Collen, P., & Parker, H. (2005). Ecological impact of the beavers Castor fiber and Castor canadensis and their ability to modify ecosystems. *Mammal Review, 35*(3–4), 248–276.
22. Grover, A. M., & Baldassarre, G. A. (1995). Bird species richness within beaver ponds in south-central New York. *Wetlands, 15*(2), 108–118.
23. Benson Ayers, M. (1997). *Aerial multispectral videography for vegetation mapping and assessment of beaver distribution within selected riparian areas of the Lake Tahoe Basin* (p. 71). Thesis, University of Nevada at Reno. Retrieved February 10, 2015.
24. Pollock, M. M., Beechie, T. J., & Jordan, C. E. (2007). Geomorphic changes upstream of beaver dams in Bridge Creek, an incised stream channel in the interior Columbia River basin, eastern Oregon. *Earth Surface Processes and Landforms, 32*(32), 1174–1185.
25. Demmer, R., & Beschta, R. L. (2008). Recent history (1988–2004) of beaver dams along Bridge Creek in Central Oregon. *Northwest Science, 82*(4), 309–318.
26. Nowak, R. M. (1991). *Walker's mammals of the world* (Vol. I, 5th ed., pp. 364–367). Baltimore, MD: Johns Hopkins University Press.
27. Seton-Thompson, cited in Sun, L., Müller-Schwarze, D. (2003). *The beaver; Natural history of a Wetlands Engineer* (pp. 97–98). Ithaca, NY: Cornell University Press.
28. Arner, D. H., & Dubose, J. S. (1982). The impact of the beaver on the environment and economics in the southeastern United States. *Proceedings of International Wildlife Conferences, 14*, 241–247.
29. Retrieved February 11, 2015, from http://icwdm.org/handbook/rodents/beavers.asp
30. Retrieved from http://www.ehow.com/about_6128827_plants-animals-tierra-del-fuego.html
31. Retrieved from http://www.nytimes.com/2014/09/26/world/americas/in-chile-a-fight-over--guanaco-hunting.html?_r=0
32. Cussac, V., Ortubay, S., Iglesias, G., Milando, D., Lattuca, M. E., Barriga, J. P., Battini, M., & Gross, M. (2004). The distribution of South American galaxiid fishes: The role of biological traits and post-glacial history. *Journal of Biogeography, 31*, 103–121.
33. Anderson, C. B., Martinez Pastur, G., Lencinas, M. V., Wallem, P. K., Moorman, M. C., & Rosemond, A. D. (2009). Do introduced North American beavers (*Castor canadensis*) engineer differently in southern South America? An overview with implications for restoration. *Mammal Review, 39*(1), 33–52.

34. Retrieved February 13, 2015, from http://www.nature.com/news/2008/080618/full/453968a.html
35. Choi, C. (2008). Tierra del Fuego: The beavers must die. *Nature, 453*(7198), 968.
36. Malmierca, L., Menvielle, M. F., Ramadori, D., Saavedra, B., Saunders, A., Soto Volkart, N., Schiavini, A. (2011). Eradication of beaver (Castor canadensis), an ecosystem engineer and threat to southern Patagonia. In: C. R. Veitch, M. N. Clout, & D. R. Towns (Eds.) *Island invasives: Eradication and management. Proceedings of the International Conference on Island Invasives* (pp. 87–90). Gland, Switzerland: IUCN and Auckland, New Zealand: CBB. xii + 542pp.
37. Valenzuela, A. E. J., Anderson, C. B., Fasola, L., & Cabello, J. L. (2014). Linking invasive exotic vertebrates and their ecosystem impacts in Tierra del Fuego to test theory and determine action. *Acta Oecologica, 54*, 110–118.
38. Ulloa, E., Anderson, C. B., Ardón, M. L., Murcia, S., & Valenzuela, A. E. J. (2012). Organic matter characterization and decomposition dynamics in sub-Antarctic streams impacted by invasive beavers. *Latin American Journal of Aquatic Research, 40*(4), 881–892.
39. Anderson, C. B., & Rosemond, A. D. (2007). Ecosystem engineering by invasive exotic beavers reduces in-stream diversity and enhances ecosystem function in Cape Horn, Chile. *Oecologia, 154*, 141e153.
40. Anderson, C. B., & Rosemond, A. D. (2010). Beaver invasion alters terrestrial subsidies to subantarctic stream food webs. *Hydrobiologia, 652*, 349–361.
41. Moorman, M. C., Eggleston, D. B., Anderson, C. B., Mansilla, A., & Szejner, P. (2009). Implications of beaver Castor canadensis and trout introductions on Native fish in the Cape Horn Biosphere Reserve, Chile. *Transactions of the American Fisheries Society, 138*, 306–313.
42. Martínez Pastur, G., Lencinas, M. V., Escobar, J., Quiroga, P., Malmierca, L., & Lizarralde, M. (2006). Understorey succession in Nothofagus forests in Tierra del Fuego (Argentina) affected by Castor canadensis. *Applied Vegetation Science, 9*, 143–154.
43. Anderson, C. B., Lencinas, M. V., Wallem, P., Valenzuela, A., Simanonok, M., & Martínez Pastur, G. (2014). Engineering by an invasive species alters landscape-level ecosystem function, but does not affect biodiversity in freshwater systems. *Diversity and Distributions, 20*(2), 214–222.
44. Parkes, J. P., Paulson, J., Donlan, C. J., & Campbell, K. (2008). *Estudio de factibildad deerradicar el castor Americano (Castor canadensis) en La Patagonia*. Landcare Research Contract Report LC0708/84.

African "Killer Bees" in the Americas

When bees attacked a Texas couple over the weekend, it seemed like a scene out of a horror film. More than 30,000 bees swarmed the couple as they exercised their small horses. At first, the woman and one of the horses jumped into the pool to avoid the stings. There were so many bees that the sky turned dark, as if it was night, she said.
The couple eventually escaped into the safety of their house, covered in bee stings, but the horses suffered a fate much worse. The first horse died before help arrived; the second died several hours later despite the efforts of an equestrian veterinarian. Five hens also died in the attack (From Associated Press)

The Honey Bee

The honey bee (*Apis mellifera*) is one of the most well-known social insects. In their colonies, encompassing from several hundred to tens of thousand "workers" (sterile females), one queen, and tens to hundreds of fertile males, they collect and protect the eggs of the queen, feed the larvae, and collect honey as a supply for winter or other hard times without nectar-containing flowers.

At intervals a new queen will appear, despite the attempts of the established one to prevent it, and the colony will split, when a queen with her entourage flies off and swarms to find another colony.

A well-known feature of honey bees is the dance that individuals returning to the hive with nectar and pollen perform to tell the others where the treasures are to be found.

Nectar, as it's present in the flower, contains a mixture of fairly complex sugars and some 80% water. Enzymes in the stomach of the collecting bee will start the process of breaking down the more complex sugars to simpler forms. Back in the hive, the collector bee regurgitates the already modified nectar, which is sucked up by younger hide-dwelling bees and further processed in their stomachs. When they in turn regurgitate it into honeycomb cells of the wax cakes, the sugars are in simple, readily utilizable forms, and the water content is lower. Further evaporation of water, speeded up by the hive bees, who flipper their wings over the not-yet-closed

A. Jernelöv, *The Long-Term Fate of Invasive Species*,
DOI 10.1007/978-3-319-55396-2_17

honeycomb cells, leads to a final water content of only about 14% in the honey. With the resulting high sugar concentration, the honey is effectively preserved.

The honey bee has a natural distribution over Europe, western Asia, and parts of Africa. Within its range of distribution, at least 20 subspecies have been characterized and named, e.g., *A.m. mellifera* "the German bee," with a natural distribution from the British Isles to Central Eastern Europe; *A.m. ligustica* in Italy; and *A.m. scutellata* in Africa.

Honey bee (*Apis mellifera*) collecting nectar (photo from FreeNaturePictures.com)

The African Honey Bee and Beekeeping in Southern Africa

In comparison to the European subspecies, the behavior of the African honey bee differs in a number of small but significant ways. Firstly, when foraging they focus relatively more on pollen than on nectar. By and large, pollen is more important as feed for the emerging brood, while nectar is more for the adults and, when converted to honey, the feed reserves for the colony in lean times.

This, of course, can be seen as an adaptation to a longer foraging break in winter for the European subspecies. The difference may also be coupled to the age-dependent task distribution within a bee colony. The younger bees have in-hive duties focusing on caring for the queen, eggs, and larvae, while the older ones are foraging for nectar, pollen, and wax. The African bees normally have a larger part of the adult population in the hive than the European ones. Whether this is due to a different age for change of duties or to a shorter lifespan is debated, but the result is a different balance between reproduction on one hand and building of reserves for hard times to come on the other [1].

The African bee will reproduce faster, swarm more often, and tend to relocate rather than tuff it out when conditions become severe [2]. When they perceive a threat to the hive, African bees send out three to four times as many defenders as their European cousins, and they pursue the intruder for a much longer distance.

Beekeeping has a long history in Africa, where traditionally the beehives were hollowed out tree trunks or large bark flakes rolled into cylinders, with as small an entrance hole as possible to make it easier for the bees to defend it. The volume was mostly small, seldom exceeding 50 l. They would typically be hung several meters up in a shady place. The honey was among other things used to make a low-alcohol "beer."

Honey Bees in America

Honey bees are not native to the Americas. They were imported from Europe and brought to both North and South America at many occasions from the seventeenth century and onward. In North America, where the first probably arrived in Virginia in 1622, they did well on balance, and there are managed beehives and wild bees of European origin in agricultural areas all over the continent. Over time, the original German bee (*Apis mellifera mellifera*) was crossbred with and later for all practical purposes replaced by the Italian bee (*A.m. ligustica*).

In the subtropical and tropical South America, the European bees didn't fare so well, and colonies there had to be given special care in order to survive and remain productive.

Introduction of African Bees to the Americas

In the 1950s Brazilian entomologists working with bees learned from a publication by Smith [3] that bees from Africa had some desired properties that the European honey bee, which had been imported from Italy a century earlier, was lacking. One of them, Warwick Kerr, imported fertilized queens of African bees (*Apis mellifera scutellata*) from South Africa. As the two types of bees could crossbreed, he sat out to do just that in order to produce a bee that incorporated an intensive grooming urge, (which makes them less susceptible to parasites like mites) with more industrious nectar foraging (leading to more honey production). As honey theft was also a problem for Brazilian beekeepers, the fact that the African bees defend their hives more aggressively than the Italian ones was also seen as a plus. The crossbreed has even been given a Latin subspecies name of its own (*A.m. adansonii*).

However, in 1957, hybridized as well as African bees escaped and formed self-sustained populations. That happened in south Brazil, in the state of Minas Gerais, not far from Uberlândia. Sometime after, it was noted that this new bees were far more aggressive than the European bees, which, up till then, were the workers in the honey production and plant fertilizing business in Brazil [4].

"Killer Bees": The Role of Generals and the Media

In 1964 a military "coop d'état" took place in Brazil and generals formed a military government. Warwick Kerr was a militant and load-voiced opponent. Being a world-renowned geneticist, the generals hesitated to deal with him in the same way as they dealt with other critics. Probably in order to destroy his credentials, the regime started to encourage media to report on the aggressive bees and link his name to every report.

The Brazilian media played along, and a flood wave of reports appeared about people and cattle being attacked and killed by African or Africanized bees that came at them without provocation.

In 1965 the US press caught on. *Time* magazine, in its September 24 issue of that year, recapped the reports from the Brazilian press and introduced the term "killer bees." Other US media followed as well as press all over Europe and South America [5].

The killer bee was born, and the media went on to produce a widespread scare for the new "Frankenstein's monster." In the Brazilian military regime and the media's retake of this international press interest in the subject, Warwick Kerr was clearly, obviously, identified as Frankenstein, and the "killer bee" as his monster.

The type of pictures of bee-covered humans that often accompanied the "killer bee" stories (photo from http://www.bing.com/images/search?q=Killer+Bees&FORM=RSRIMG#view=detail&id=608B939E4DECB408B5D7FAF0C1B57BA5D26FFE90&selectedIndex=25)

The Spread of African and Africanized Bees

After the release in 1957 in the Brazilian state of Minas Gerais, the African and Africanized bees spread in all directions. Parallel to the spread of the runaway bees and their offspring, the hybrid bees of Dr Kerr were also introduced to beekeepers and passed on between them, providing a human-supported distribution pathway.

As these bees of African origin swarmed often and simply abandoned hives if they found the area unattractive, beekeepers often lost them. However, three characteristics that made them different from their European cousins soon became well known to rural Brazilians: their aggressiveness, their sturdiness, and the fact that they were reliable honey producers. Poor rural people also picked up the habit of their African brothers, to manufacture simple makeshift hives and put them out on trees in shady places, finding that the probability was good that a bee swarm sooner or later would find and occupy them.

So, altering between a life in the wild and as "domesticated" producers of honey for human consumption, the bees spread on a combination of on their own account and through human involvement. To start with, this happened without much ado, but from 1964 and onward was increasingly accompanied by media reports of the menace of the killer bee and stories of livestock and human fatalities. By that time they had reached Sao Paulo and continued their spread with an estimated average speed of 320 km per year. In 1966 the Africanized bees reached Paraguay and Argentina. The Amazon and the Andes slowed down the pace, but were in no way insurmountable, and in 1976 they reached Venezuela [6], in 1982 Panama [7], and in October 1986 Mexico [8].

In October 1990 they were first found in the USA, near Hidalgo, Texas, and by the turn of the millennia, they occurred in over much of the US south: Texas, California, New Mexico, Arizona, Oklahoma, Louisiana, Arkansas, Alabama, and Florida. However, the extension northward appears climatically limited, and in North America the Africanized bees have not extended north of the 34° latitude [9].

So far, the Africanized bees invaded Trinidad and Tobago, in 1999 [10], but have not reached most of the other Caribbean islands.

During its spread these bees have interbred with the different European honey bees that were introduced to the regions earlier.

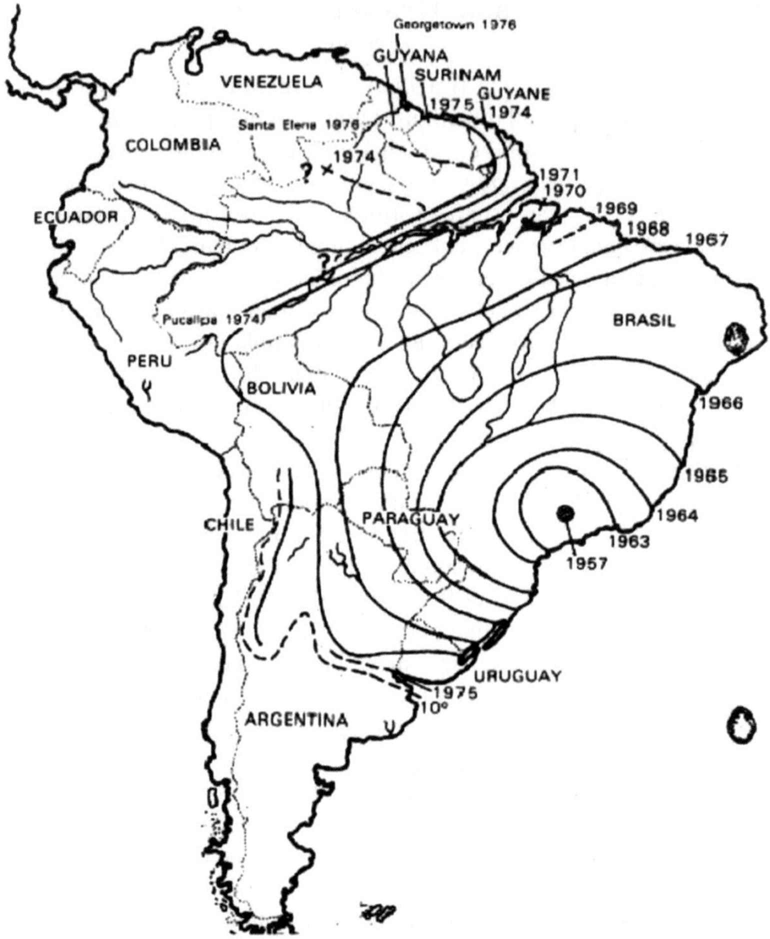

A map of the spread over South America from the release in Minas Gerais in 1957 and twenty years on (from Taylor) [11]

Ecological Effects

From an ecological point of view, the important thing with the arrival of the Africanized bees is that the total number of honey bees has increased, especially in the warmer parts of the Americas. The native bee fauna does not contain any species with the social organization, large colonies, and honey-storing habits of the honey bee. Neither are they generalists with regard to nectar and pollen collection but rather specialized on a species or families of plants. For them, the honey bee is a formidable competitor, and the number of individuals and species drops locally, when the honey bee comes on the scene [12].

With regard to plants, honey bees seem in general to be able to pollinate also the native ones. Indirectly, though, they affect them negatively by pollinating introduced plant species, a thing the specialized native bees would generally not do, thereby making the intruders competitive.

Economic Effects

In economic terms the plus factors of the Africanized bees are in honey production and pollination. With the arrival of the African and Africanized bees, Brazil went within a decade from 34th to 7th place among honey-producing countries [13]. When the "killer bees" reached other Latin-American countries, e.g., Argentina and Mexico, the honey production increased even more there, pushing Brazil somewhat back again to place 11.

If the increase in the volume of honey, produced in the first 20 years after their arrival to the different countries, is attributed entirely to the new bees, more than 200,000 tons per year would fall on their account. Much of this is exported, generating revenues of around \$500 million annually. Home market consumption may add another \$150–200 million. Obviously, also other factors have contributed to the increase, so half of these figures may be a more realistic assessment of the contribution of the Africanized bees. However, bees produce more than honey, and pollen and wax are other commercial products, but often nowadays pollination is seen as their most important contribution to the economy, boosting output of, e.g., almonds, fruits, berries, and beans. A larger number of honey bees, especially in the tropics and subtropics of Latin America, as is a result of the arrival of the Africanized bees, will have pushed pollination too.

In the USA, the value of honey bee pollination is assessed to be in the range of \$20–30 billion [14], while the value of honey production is \$320 million [15] (Figures for 2013). Thus, the value of pollination is 60–90 times that of honey.

If one were to use a similar multiplication factor for Latin America, the African "killer bees" would have contributed an astonishing \$25 billion per year in pollination services.

Dr Kerr, despite being demonized for introducing the bees from Africa, would probably have seen this as a roaring success.

On the negative side are damages to livestock and people caused by aggressive bees. This is hard to evaluate in economic terms, and the exaggerated reporting in media doesn't make an assessment any easier, but it seems clear that the economic value of the damage stays well below the \$100 million per year mark and thus would be small compared to the advantages.

Control Attempts

In the 1970s and 1980s, when the advance of the African bees was in full swing and the media reports scare-mongered stories about the devastation that might lay ahead, if they entered and spread in the USA, American-financed attempts were made to halt them [16]. In Panama as well as in Mexico, swarm capture (and kill) programs were tried. Mostly, bait hives were used and sprayed as they became occupied [17]. Once the "killer bee" had reached Texas, these efforts to prevent further spread were intensified and millions of dollars spent on various programs—all in vain.

Among the methods proposed, but officially not tested, was to use a bee from the southmost tip of Africa, *A.m. capensis*, that had been shown to exert a sort of "social parasitism" on *A.m. scutellata*. The workers of a specific strain of this subspecies can produce a specific pheromone, attain reproductive status, and overthrow the *A.m.s.* queen [18].

In 1990, in South Africa, *A.m.c.* colonies were brought into proximity of those of *A.m.s.* Ten years later, by subverting worker recognition and queen regulation, a single linage of the *A.m.c.* was devastating *A.m.s.* colonies all over northern South Africa [19].

A reason to remain open for the possibility that some traits from the *Apis mellifera capensis* subspecies could have reached the Americas is a behavior of what's called "the African bee" in a report from the University of Florida [11]:

> A behavioral curiosity of African bees concerns nest usurpation (or colony takeover) of European colonies. Small African swarms containing a queen often land on the outside infrastructure of a European colony (a wall, beekeeper-managed hive, etc.). As time passes, the worker bees in the African swarm begin to exchange food/pheromones with the European workers from the colony. This gradually ensures the adoption of the African bees into the European colony. Somewhere during this process, the European queen is lost (perhaps killed by the African bees; her fate remains uncertain at this point) and the African queen is introduced into the colony, thus becoming the reigning matriarch. European bees do not display this behavior but often fall victim to it, thus creating an African colony from a preexisting European one.

Current Situation

Today, the Africanized bees can be said to have concurred the Americas between 32° south and 34° north with the exception of high mountain ranges in the Andes. It can also be said that the African bees have drowned in the populations of European honey bees in the Americas and that, by now, what remains of them are only some genetic traits in the overall gene pool of the honey bee. Both points of view are in a certain meaning correct. Over its range of distribution, the honey bees in the North and South Americas show no specific African traits, that's why one can say that the above latitude limits apply. In the tropical and subtropical parts of the continent, though, African traits are clearly visible, even dominating, and in the areas with a climate in between, some are. Genetic selection within the species, based on the added diversity that the African bees brought, is working toward a regional optimization and differentiation.

For beekeepers in the warmer parts of the Americas, this has meant more productive bees. No doubt, the new honey bees are more aggressive than the old ones, but Latin-American keepers have learned how to handle them, and human causalities are very few.

Public Perception and Official Policies

For part of the general public, especially those living in urban areas and the parts of North America to which the Africanized bees have not yet arrived, the media picture of the "killer bees" is still alive. For much of the Latin-American rural population and especially where beekeeping is practiced, they are accepted as revenue-generating creatures that might occasionally sting.

In the USA the official policy is still to limit their spread and if possible push them back. Significant efforts toward this goal are continuously made. In Latin America most countries have changed their negative attitude to one of acceptance in light of the economic benefits, and also where the earlier restrictive policies are still in the books, they are seldom acted upon.

Summary

The "killer bee" was to a large extent a media hype, grossly exaggerating the aggressiveness and danger of the African and Africanized bees as they spread over the Americas. These bees are more aggressive than their European cousins, but not more so than that trained beekeepers can deal with. Furthermore, in the tropics and subtropics, they are more productive than the European subspecies, and they are generally less affected by parasites and diseases. Since introduction in Brazil in 1956, the African bees have interbred with European ones both spontaneously and through human interference. As a result, the honey bee in the Americas has gotten a higher genetic diversity with some African traits. Natural (and human) selection has favored some of those in the tropical and subtropical areas but reduced the frequency of them in temperate climates. One could justify both the statement that the Africanized bee has taken over most of the Americas and that it has drowned in the overall honey bee gene pool. Economically, the Africanized bees have had a positive impact, ecologically a negative, but not very strong, one.

References

1. Fewell, J. H., & Bertram, S. M. (2002). Evidence for genetic variation in worker task performance by African and European honeybees. *Behavioral Ecology and Sociobiology, 52*, 318–325.
2. Winston, M. L., Taylor, O. R., & Otis, G. W. (1983). Some differences between temperate European and tropical African and South American honeybees. *Bee World, 64*, 12–21.
3. Smith, G. (1953). Beekeeping in the tropics. *Bee World, 34*, 233–245.
4. Winston, M. L. (1992). *Killer bees: The Africanized bee in the Americas*. Cambridge, MA: Harvard University Press.
5. Dr Kerr and the killer bee. http://www.badbeekeeping.com/kerr.htm
6. Thimann, R. (2001). 25 Years of the Africanized honey bee in Venezuela. *Proceedings of the 37th International Apicultural Congress*, 28 Oct–1 Nov 2001, Durban, South Africa, Apimondia. http://www.apimondia.com/congresses/2001/Papers/275.pdf
7. Globendia. *La abeja asesina en serie (africana) anida ya en nuestra árbles. Es muy agresiva.* http://globedia.com/abeja-asesina-serie-africana-arboles
8. *Africanized honey bees*. Utah County Bee Keepers Association. http://utahcountybeekeepers.org/africanized_honey_bees.html
9. Castillo, K. (2010, October 27). Africanized bees make T&T honey so sweet. *Trinidad Express Newspaper*.
10. Taylor, O. R. (1977). Past and possible future spread of Africanized honey bees in the Americas. *Bee World, 58*, 19–30.
11. *African honey bee, Featured creatures*. IFAS, University of Florida, http://entnemdept.ufl.edu/creatures/misc/bees/ahb.htm
12. Genaro, J. A. (2006). A history of systematic studies of bees of Cuba (Insecta, Hymenoptera, Anthophila). *Zootaxa, 1195*, 39–60.
13. Mac Margolis "Market Scene: 'Killer' Bees Make Honey of a Deal with Brazil: Production has soured since the ornery invaders took over beekeepers' hives", April 5, 1994, *Los Angeles Times*
14. USDA. (2013). Report on the National Stakeholders Conference on Honey Bee Health. Retrieved from http://www.usda.gov/documents/ReportHoneyBeeHealth.pdf
15. http://www.statista.com/statistics/191993/value-of-honey-production-in-the-us-since-2000/
16. Morse, R. A. *Killer bees: Origin of the name and other humbugs*. http://ageconsearch.umn.edu/bitstream/130802/2/RogerMorse.pdf
17. Seeley, T. D., & Morse, R. A. (1982). *Bait hives for honey bees* (6pp). Cornell Cooperative Extension Publication, Information Bulletin No. 187.
18. Schmid-Hempel, P. (1998). *Parasites in social insects*. Princeton, NJ: Princeton University Press.
19. Hartel, S., Neumann, P., Raassen, F. S., Moritz, R. F. A., & Hepburn, H. R. (2006). Social parasitism by Cape honeybee workers in colonies of their own subspecies (Apis mellifera capensis Esch.) *Insectes Sociaux, 53*, 183–193.

The Warty Comb Jelly in the Black Sea

One of the most spectacular collapses in fisheries yields occurred in the Black Sea in the late 1980s. In the mid-1980s, the bulk of fish biomass and catch there was made up of small, plankton-eating pelagic fish such as anchovy, sprat, and horse mackerel. Five years later, catches in the territorial waters of Russia, Ukraine, and Georgia were down by some 90% overall, as illustrated by the diagram below. The trends in figures from Turkish, Bulgarian, and Romanian waters were similar for the period 1984–1991.

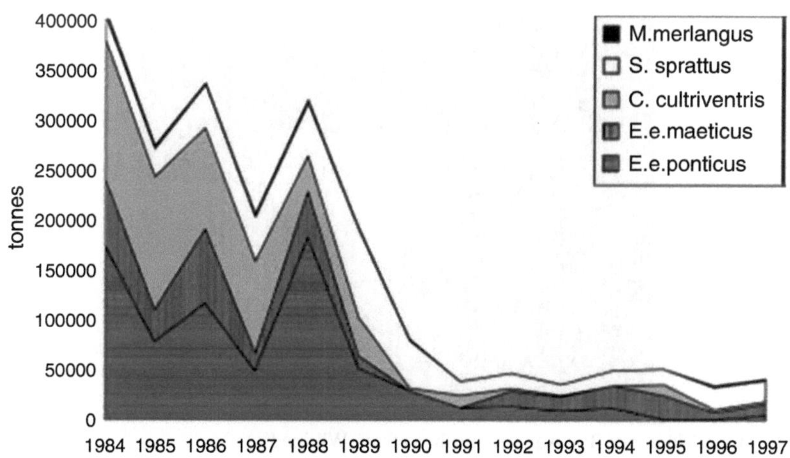

Catches of zooplanktivorous species in the Black Sea (1984–1977) (from FAO Statistical Bulletin)

The reason was the arrival and population explosion of the invasive warty comb jellyfish, *Mnemiopsis leidyi*, from its native home in the western Atlantic waters.

A. Jernelöv, *The Long-Term Fate of Invasive Species*,
DOI 10.1007/978-3-319-55396-2_18

Biology and Original Zoogeography of the Warty Comb Jelly

Mnemiopsis leidyi is a species of tentaculate ctenophore native to western Atlantic coastal waters stretching from the Canadian Maritime Provinces to southern Argentina. It is bell-shaped and transparent and measures some 10 cm in length with a diameter of 2.5 cm. It has several feeding tentacles and rows of ciliated combs that run vertically along the body and that glow blue-green when the animal is stressed. The warty comb jelly is blind and brainless, does not sting, and moves so slowly that its nickname is "sea walnut." It tolerates a wide range of salinities from 4 to 38 psu, temperatures from 4 to 32 °C, low oxygen, and fairly high levels of pollution. Hermaphroditic, it can self-fertilize: sperm and eggs are released into the water, where fertilization takes place. Egg production starts when the animals reach about 15 mm in length. A large specimen can produce 14,000 eggs in a day [1]. *Mnemiopsis* consumes plankton, especially crustaceans; eggs and larvae of fishes and mollusks; jellyfish; and other comb jellies, including smaller individuals of its own kind. When food is abundant, it can eat ten times its own body weight in a day [2].

It is in turn preyed upon by other ctenophores such as *Beroe*, jellyfish of the *Scyphozoa* group, and several species of birds and fish.

In nutrient- and plankton-rich coastal lagoons in its native range, the warty comb jelly can sometimes, temporarily, reach high densities and affect zooplankton population size, but this has been too local and temporary to be seen as a problem [3].

M. leidyi at the Monterey Bay Aquarium (photo from Wikipedia)

The Black Sea

The Black Sea is a landlocked body of water with a surface area of 436,400 km^2, five times that of Lake Superior, and a volume of 547,000 km^3, 50 times that of Lake Superior. It is meromictic, meaning that it is permanently stratified, and has three distinct water layers. At the bottom is the deepwater layer, encompassing some 90% of the sea's water volume, which is more saline than the surface water, is anoxic with a high concentration of hydrogen sulfide, and has a year-round temperature of almost 9 °C. This layer is thus toxic to most higher organisms. Above the bottom water layer are two further layers that mix off and on in the winter, when their temperatures are equal. Normally, the lower of these, called the cold intermediate layer (CIL), is colder and thereby denser, than the surface water, but with more or less the same salinity. The border between the bottom layer of water and CIL is at between 100 and 200 m depth and that between CIL and the surface water layer lies at 50–100 m. The salinity of the surface water varies between only a few psu outside of the mouths of large rivers such as the Danube, the Don, and the Dnieper and 17 psu where water leaves the Black Sea for the Mediterranean at the Bosphorus. Temperatures range from zero to up in the vicinity of 30 °C, depending on the place and the time of the year.

The nutrient balance, of the Black Sea is largely dependent on the influx of nutrients via the rivers and some recirculation from the bottoms of shallow areas, as the permanent stratification means that nutrients from organic matter that settle into the bottom water are trapped there.

The Black Sea (picture from Wikipedia)

After World War II, the use of fertilizers in agriculture in areas prone to runoff led to an order of magnitude increase in nutrient input into the sea [4] and to subsequent eutrophication with high phyto- and zooplankton production that also increased the standing crop of small zooplankton-eating fish. During the same period, intensive industrial fishing for larger fish such as sturgeon, bonito, blue fish, and mackerel made these fish rare, in some cases pushing them to the brink of extinction [5].

Bottom trawling, one of the new fishing practices introduced around that time, also had the effect of mechanically destroying bottom biotopes and damaging benthic communities. A further deleterious effect on the ecosystem was caused by bottom sediments (pelitic fractions) being suspended into the pelagic layer. This induced secondary eutrophication, decline of the sea's transparency, and depression of photosynthesis processes of bottom-living plants. However, the main negative effect of bottom trawling was recognized to be the silting of vast bottom areas [6].

In addition, three previously common marine mammals, the Black Sea bottlenose dolphin, *Tursiops truncatus ponticus*; the Black Sea common dolphin, Delphinus *delphis ponticus*; and the Black Sea harbour porpoise, *Phocoena phocoena relicta*, saw substantial population declines. This meant another boost to those small fish species from less predation and reduced food competition.

The rivers did not only carry an increased load of nutrients to the Black Sea; the quantities and concentrations of a wide range of pollutants also increased manifold. In addition, enormous dams were built, changing the rivers' seasonal flow patterns. This, together with the disruptive effects of invasive new species such as the crustacean *Rhithropanopeus harrisii*; the mollusks *Mya arenaria*, *Rapana thomasiana*, and *Cunearca cornea*; and the fish *Mogul soiuy*, had already had profound impacts on the Black Sea ecosystem by the late 1970s. Some specific examples are the declines of the previously huge stands of macrophytes in the deltas of the major rivers and the unique, originally more than half a million ha large, attached red algae-dominated area in northwest Black Sea called Zernov's *Phyllophora* field. The biomass of the *Phyllophora* stock, that by most estimates numbered close to ten million tons in the 1960s, at present does not even reach one million tons. Reduced water transparency is considered to be one of the main reasons for this decline [7].

Another example is the massive die-off of *Mytilus* mussels following the anoxia that also started to occur in shallow bottom areas, caused by the nutrient-triggered algal blooms and increased siltation. An estimated 35 million tons of mussel was lost and with them a very significant filtering capacity, which served to reinforce the plankton explosion [8].

The Sea of Azov

The Sea of Azov can be seen as an enclosed bay in the northeastern part of the Black Sea. It is connected to the main body of water via the 4 km-wide strait of Kerch that lies between the Crimean Peninsula and the Russian mainland to the east. It has a surface area of 39,000 km^2 and is very shallow, with a maximum depth of 14 m. The current vertical profile of the Sea of Azov exhibits oxygenated surface waters and anoxic bottom waters, with the anoxic waters forming a layer of 0.5–4 m in thickness.

Two big rivers, the Don and the Kuban, and some twenty smaller ones flow into it, which means about 38 km^3 of freshwater, about the same amount as the saline water that enters from the Black Sea. As a result, water salinity is low, 10–12 psu in the open sea, and practically freshwater at the rivers' mouths. The rivers also carry huge volumes of silt, sand, and shells that settle, making the sea bottom smooth and creating numerous bays, limans (a type of delta), and narrow sand spits. The large volume of settling matters gradually makes the lake more and more shallow—less than 100 years ago, its maximum depth was over 16 m. The rivers also carry many nutrients with them, which make the Sea of Azov eutrophic and fertile.

Winter temperatures are low, and large stretches of the sea are regularly frozen.

An icebreaker on the Sea of Azov (from Wikipedia)

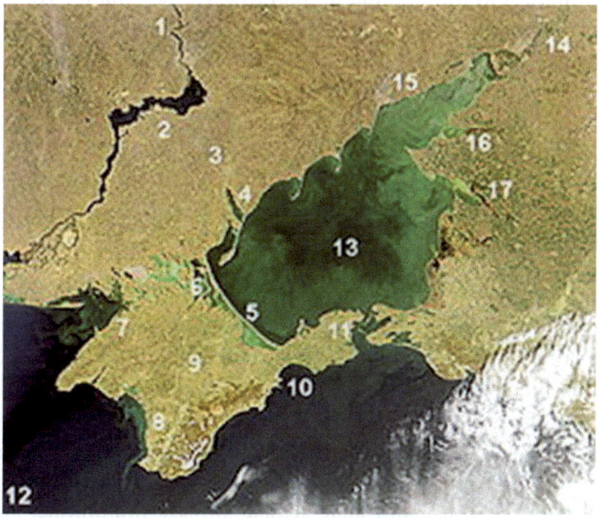

Satellite picture of the Sea of Azov (from Wikipedia)

Along the rim of the Sea of Azov, there are numerous estuaries and marshes dominated by the common reed (*Phragmites australis*), sedges (*Carex* spp.), cattail (*Typha* spp.), and bur reed (*Sparganium* spp.). The submerged vegetation is mostly limited to a depth of less than a meter, dependent on the high turbidity caused by silt and plankton algae. Many of the typical plants go under the popular name of pondweed despite belonging to different families such as *Chara* and *Nitella*, *Ceratophyllum*, and *Potamogeton*. Others are water lilies (*Nymphaeaceae)*. Two immigrants that dominate certain plant communities are the Indian lotus (*Nelumbo nucifera*), that came via Africa, and the Canadian waterweed (*Elodea canadensis*), that came via estern Europe [5]. Their habitats are the "almost freshwater lagoons" at the mouths of rivers, although starting in the late 1970s, the stands of Canadian waterweed thinned out, and it is in no way a dominating species any longer.

The Sea of Azov is very rich in phytoplankton, both in biomass and in species richness. About 600 species of planktonic algae are known to be in it. Diatoms dominate in winter and green algae pretty much the rest of the year, but blue-green algae and pyrophytes are also significant [6].

The zooplankton community is also rich, with densities often over 100 g/m^3 and sometimes up to 200 g/m^3 [9]. Copepods and Rotifers dominate, but in the least-saline parts also, Cladocera occur in high densities. In certain parts of the year, the larvae of snails, mussels, and polychaetes are also abundant. The bottom fauna biomass is dominated by mollusks [6].

Very rich in fish, typical annual catches in the Sea of Azov used to correspond to 80 kg per ha of surface area, while the figure for the Black Sea overall was 2 kg (and 0.5 kg for the Mediterranean). Traditionally, the fish, in order of the greatest commercial value, used to be anchovies, perch, sturgeon, bream, whitefish, herring, plaice, carp, mackerel, and mullet [9].

Overfishing, water pollution, and changed hydrological regimes following river damming and water withdrawal had before the 1980s caused a shift in fish populations away from the larger and more valuable fish toward the smaller zooplankton-eating ones such as anchovy, horse mackerel, and the Black Sea sprat, locally called kilka. Thus, even as the volume of the catch stayed high, its value had dropped significantly.

The Introduction, Spread, and Population Explosion of the Warty Comb Jelly

The first few individuals of *Mnemiopsis leidyi* seen in the Black Sea were sampled by scuba divers in autumn of 1982 [10], in near-shore waters off of southeast Crimea. It was assumed that the species had arrived a year or two earlier, in all likelihood with ballast water from the US East Coast. The next report came from Bulgaria, where a few specimens were net sampled near the shore in the spring of 1986. In 1987, the warty comb jelly was found in several different places in the Black Sea and in larger numbers [8]. The following year, it was already ubiquitous, and in 1989 the population exploded and reached a density of some 400 specimens/m^3 and an estimated biomass of a billion tons in wet weight. In the open sea, the wet weight/m^2 of surface area was 4.6 kg and was even much higher in some coastal lagoons.

The very high jelly density figures continued into the spring of 1990, but by summer they started to decrease. Assessments and ecosystem modeling suggest that their density in 1989–1990 represented an overshoot and that those numbers were not sustainable. After a few years with a downward population trend, the *M. leidyi* numbers started to increase again in 1994, peaked again in 1995, and again followed by a decrease for some years. In 1998 a third peak followed, with average open-sea densities of 876 g/m^2 [10].

In the Azov Sea, the warty comb jelly was first found in August 1988. As it cannot survive the cold winter waters and does not reproduce in the low-salinity waters there, every year it has to be repopulated from the Black Sea. After the penetration of *M. leidyi* into the Sea of Azov, which happens in April–July through the strait of Kerch on the incoming salty current, the individual jellies grow fast and spread throughout the sea within a week. The densities in the Sea of Azov are thus dependent on the densities in the northern reaches of the Black Sea, but they are also influenced by the time available for growth before winter sets in. Sometimes, as in 1993, if the autumn weather is good, even a late-arriving jelly population can grow to have a high biomass. That year, warty comb biomass reached 20 million tons by late November, not too far from the record of 30 million tons in 1989, a year in which they had arrived in large numbers already in April, although they started to die off in early October.

Ecological Effects of the *M. leidyi* Invasion from Late 1980 to Late 1990

The warty comb jelly arrived to the Black Sea that provided almost ideal conditions for it. A number of factors, some of them listed above, contributed to this Schlaraffenland—this "land of plenty." With plenty of food and no predators, the jelly population exploded. The year 1989, however, represented an overshoot, and it became apparent that that size of jelly population was not sustainable. Before it fell back to somewhat less staggering levels in the following years, however, the warty comb jelly wreaked havoc on the entire ecosystem and pelagic, in particular, the zooplankton-eating fish. These were hit in two ways: the warty comb jelly both outcompeted them for food and consumed their eggs and fry.

Zooplankton

The 1989 mass development of *M. leidyi* triggered an unprecedented decline in density, biomass, and species richness of mesozooplankton. That summer, the copepods *Paracalanus parvus* and *Centropages ponticus* almost disappeared. *Oithona nana* and the *Potellidae* species were not found in any samples. The standing stock of *Oithona similis* and *Acartia clausi* all but collapsed, as did many other, normally common, zooplankton, such as cladocerans, appendicularians, and the larvae of gastropods and polychaetes. A planktonic predator on zooplankton, the arrow worm, *Sagitta setosa*, also virtually disappeared [10]. Some species of zooplankton, such as the copepod *Calanus euxinus*, were not much affected in 1989, but their abundance started to drop in 1990.

In the following years, zooplankton populations recovered to some extent, as the warty comb jelly declined, but suffered severely again in 1994–1995 and in 1998, when populations of the jelly peaked again.

Jellyfishes and Jellies

There are three indigenous gelatinous zooplankton predators in the Black Sea, the barrel jellyfish (*Rhizostoma pulmo*), the moon jellyfish (*Aurelia aurita*), and the sea gooseberry (*Pleurobrachia pileus*). The sea gooseberry is generally found in polluted, eutrophic coastal waters with low salinity. The barrel jellyfish inhabits the CIL at depths from 25 to 150 m, and the moon jellyfish occupies the surface waters both in coastal areas and in the open sea. In the late 1970s and early 1980s, the population of *A. aurita* gradually increased and reached a peak, with an average density of almost a kg wet weight/m^2, or some 4×10^8 t.

The moon jellyfish is the only one of the three that is in direct competition with the warty comb jelly, as they share both habitat and feed. It was also hard hit when the competitor arrived, and its density dropped to some 20% of the peak values [7].

In 1997, a new alien species of comb jelly, *Beroe ovata*, was found in a coastal area of the northern Black Sea [11]. It is claimed that it, too, accidentally arrived with ballast water—an interesting coincidence, if so, as an expert group had shortly before suggested that it should be brought in as a biological control agent. Its arrival signaled a new chapter in the comb jelly invasion of the Black Sea.

Fish

In the 1980s, before the warty comb jelly became an important component of the Black Sea pelagic ecosystem, three fish species dominated the standing crop and catch, the Black Sea anchovy (*Engraulis encrasicolus ponticus*), the Mediterranean horse mackerel (*Trachurus mediterraneus ponticus*), and the sprat (*Sprattus sprattus phalericus*). The standing crop of them was assessed to be 2–3 million tons, of a total fish biomass of 3.5–4 million tons. The total fish catches in the Black Sea in the period 1984–1988 amounted to some 700,000 t, out of which 2/3 were caught in Turkey. The anchovy made up the bulk of the total, with a peak of some 500,000 t in 1988 [3].

Several scholars see this as overexploitation of the stock and attribute to it some of subsequent rapid decline in catch weights. In 1989, as the warty comb jelly population exploded, the catch of anchovy dropped to 64 t in the CIS countries and to below 100,000 t in Turkey. In the following year, it was less than 100,000 t combined.

On a smaller scale, the development was even more catastrophic in Romania. There, the anchovy catch was almost 2000 t in 1988 and dropped to 66 t in 1989 and to 5 t in 1990.

In the following few years, the anchovy landings rose again somewhat in Turkey (to 100,000 t) and Romania (50 t) but remained very low in the CIS countries, partly for socioeconomic reasons as fishing efforts went down in connection with the implosion of the Soviet Union.

A USSR survey conducted in the winter of 1990–1991 found an overwintering anchovy stock of only 30,000 t. The fish were lean and in bad shape, obviously half-starved from the lack of zooplankton to feed on. The anchovy population seems to have reacted to the situation in two ways, with a movement toward "premature" spawning of fish even of under a year old and in a change in its main spawning area from the northwest to the southeast Black Sea [12]. These actions may have facilitated a partial recovery of the anchovy stock in the 1990s, but only to a level of 30–50% of that in the late 1980s.

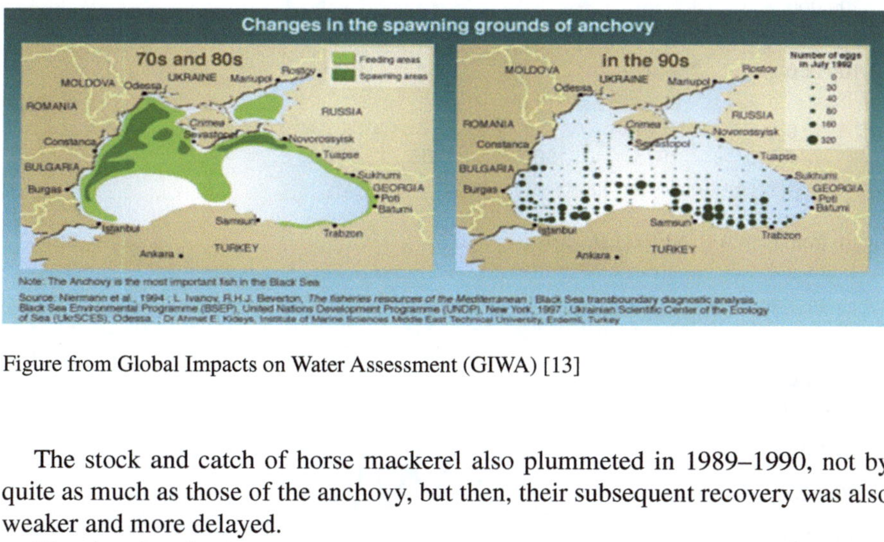

Figure from Global Impacts on Water Assessment (GIWA) [13]

The stock and catch of horse mackerel also plummeted in 1989–1990, not by quite as much as those of the anchovy, but then, their subsequent recovery was also weaker and more delayed.

The situation for the sprat was different. After an increase in population following eutrophication in the 1970s, the standing crop measured around a million tons throughout the 1980s. It then fell slowly in the first years of the 1990s to 600,000–700,000 t. The reason why the sprat fared better than the other fish was that its prime feed is the copepod *Calanus*, which spends most of its time in the CIL layer of the Black Sea, while *M. leidyi* prefers the warmer surface water [10].

In sum, overall fish catches in the Black Sea dropped from some 800,000 t in the late 1980s to a quarter of that in 1991. They subsequently recovered to some extent and hit new peaks in 1995 and 1998, of about half a million tons. However, practically the entire recovery took place in Turkish waters, with the catches in Russia, Ukraine, and Georgia remaining at a very low level. There were also changes in species composition in that, e.g., the Atlantic bonito (*Sarda sarda)*, the Atlantic mackerel (*Scomber scomber)*, the chub mackerel (*S. japonicus*), and the goat fish (*Mullus barbatus)*, a somewhat larger fish that had been rare in the Black Sea since the 1970s, all increased their share of the Turkish catches. In Russia, the regional invader, the mullet, *Mugil soleiy*, became one of the most important commercial fish [14].

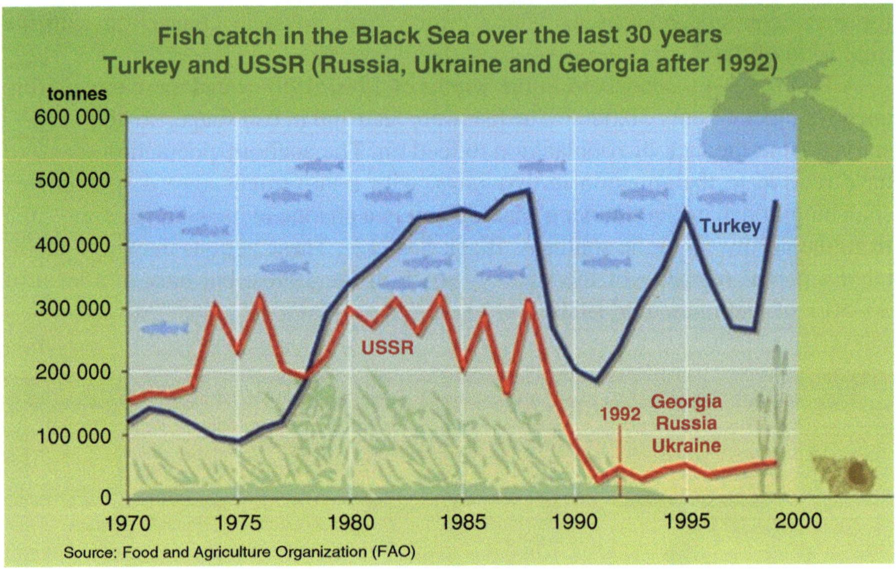

Figure from GIWA (UNEP) [13]

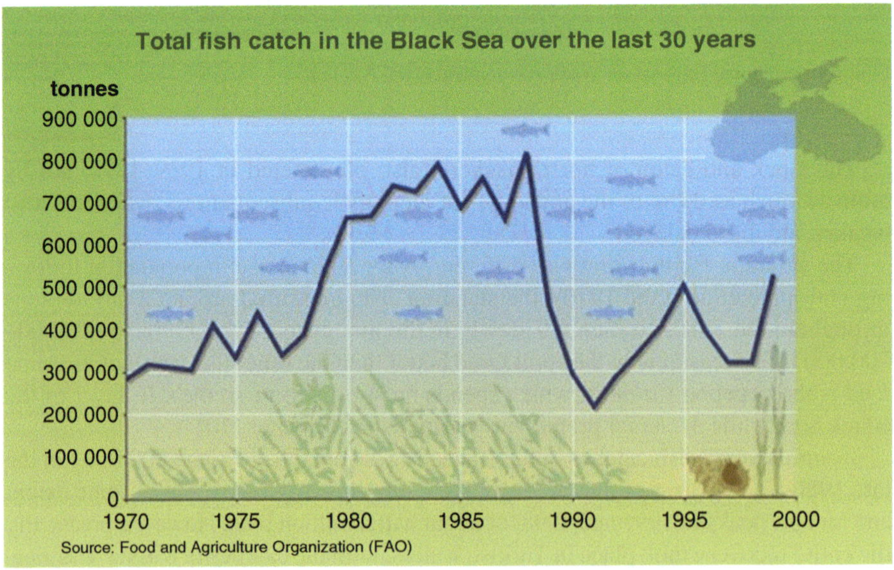

Figure from GIWA (UNEP) [13]

In the Sea of Azov, the effects of the warty comb jelly in any specific year depended very much on the time of that year's infestation. If the jelly arrives early, in April–May, it will much reduce the zooplankton population before June–July and will also prey heavily on the eggs and larvae of anchovy and sprat, undermining their reproduction. If *M. leidyi* arrives only in June–July, the small pelagic fish can spawn safely and grow to catchable sizes. Sometimes, as in 1993, if the autumn weather is good, even a late-arriving jelly population can grow to a high biomass. That year it reached 20 million tons, not too far from the 1989 record, a year in which they arrived in April.

Effects of *M. leidyi* on Nutrient Status and the Vertical Transportation of Matter

The massive development of the warty comb jelly, with some one billion tons of wet weight in 1989, and the annual die-off of most of them in the winter, lead to a quicker sedimentation of organic matter into the stagnant, deepwater layer of the Black Sea. On a larger timescale, this may not have mattered so much, but the immediate effect was an increased in hydrogen sulfide (H_2S) production from the anaerobic breakdown and settling of dead and decomposing jellies. It also meant some degree of depletion of organic matter and nutrients from the upper, oxygenated layers of the sea.

Cascading Effects

When zooplankton densities dropped, due to the predation of the warty comb jelly, the grazing pressure on phytoplankton decreased. Thus, their numbers increased and blooms became longer-lasting. The visual impression this gave was that *M. leidyi* increased eutrophication.

When the population of anchovy and other plankton-eating fish collapsed, especially in the northern Black Sea, fish-eating marine mammals such as dolphins became scarce. The population of fish-eating birds dropped as well.

The comb jelly exudes mucus from the surface of their bodies. Bacteria readily consume the mucus, and their numbers may exceed a million per ml when jellies are abundant. The bacteria plankton, in turn, is consumed by zooflagellates, dominated by *Strombidium* spp. (size 15–30 μm) and by infusoria. These groups of heterotrophic microplankton increase significantly in the presence of jellies and in proportion with their densities [15].

Beroe ovata

Like *Mnemiopsis leidyi*, *Beroe ovata* is a comb jelly, and the two share many fundamental physical characteristics. They also share much of the area of natural distribution. Among the differences are that *B. ovata* lacks tentacles, swims much faster, is bigger, and, most importantly, mainly preys on other jellyfishes and jellies, the warty comb jelly being one of them. They have modified mouth cilia that they can use to bite pieces out of their prey, if these are too large to swallow whole.

They were proposed as a possible biological control agent for *M. leidyi* in the Black Sea, by, among others, GESAMP [3], but their actual introduction is said to have been accidental, with ballast water.

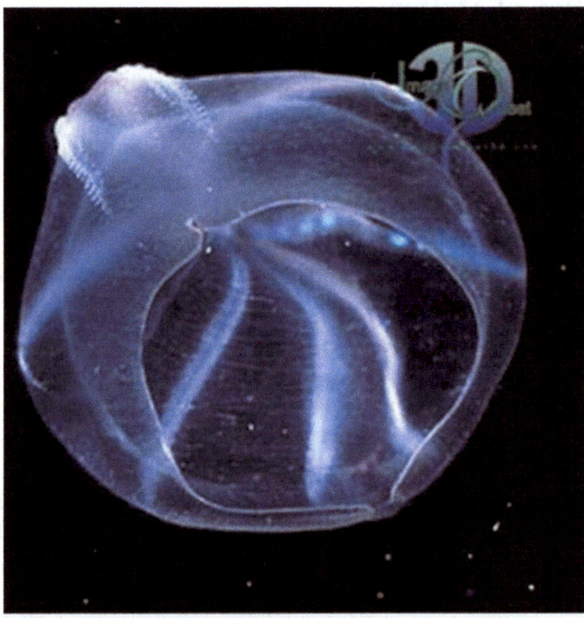

Beroe ovata, mouth agape.

Beroe ovata, mouth agape (photo from wikispaces.com)

In October 1997, *B. ovata* was first recorded in the Black Sea, in Bulgarian waters. Two years later they were common in surface waters all over the sea and seen as a new invader [16].

Their only prey in the Black Sea, besides cannibalism, was the warty comb jelly. The two species quickly developed a classical predator-prey population relationship, with a yearly, early population peak of the prey, followed by an accelerating density drop, at which time the predator population caught up [17]. Then with the prey almost gone, predator numbers would soon contract. Thus, *Beroe ovata* reached its highest densities in 1999, when it, the year before, had been able to tap into a huge supply of *M. leidyi*. In the annual cycles thereafter, it reaches 10–20% of those peak numbers, but the bloom only lasts a short time as the prey densities drop drastically.

The Warty Comb Jelly After 1997 and Consequences of the New Regime

In 1998, *M. leidyi* occurred in large numbers and with a biomass of over 500,000 million tons wet weight, short of the 1989 record of a billion tons, but not by that much. That year, the predation of *B. ovata* did not yet do much to keep the warty comb jelly at bay, but already the following year, the new regime, as described above, became established. Since then, the warty comb jelly has had only a short annual peak of at the most 2-month duration in late summer and early autumn, compared to 8–9 months in the 1990s. At their annual peak, warty comb jelly densities in the new millennium may reach 20–30% of those in the last decade of the old one.

As a consequence of the decline of the warty comb jelly, the pelagic zooplankton has recovered in density, biomass, and diversity. Pelagic fish eggs and larvae have also recovered in abundance and diversity, while phytoplankton and chlorophyll a have decreased. Among benthic groups, the density, biomass, and species richness began to rise. The stocks of pelagic planktivorous fish, such as anchovy, sprat, and horse mackerel, recovered as their diet composition and biological parameters improved. Piscivorous fish and dolphins began to make a comeback as their prey became more abundant. Fisheries registered increased catches of small pelagic fish, while other groups reappeared in the catches. [15] The authors conclude that with the advent of *Beroe ovata*, the Black Sea secured a predator that was able to control the warty comb jelly and that as a result, the sea entered a period of recovery.

Economic Effects

There are a number of different figures mentioned in the literature, of the economic effects of the warty comb jelly invasion of the Black Sea. Most of them are based on data such as those in the curves below covering the years 1975–1995.

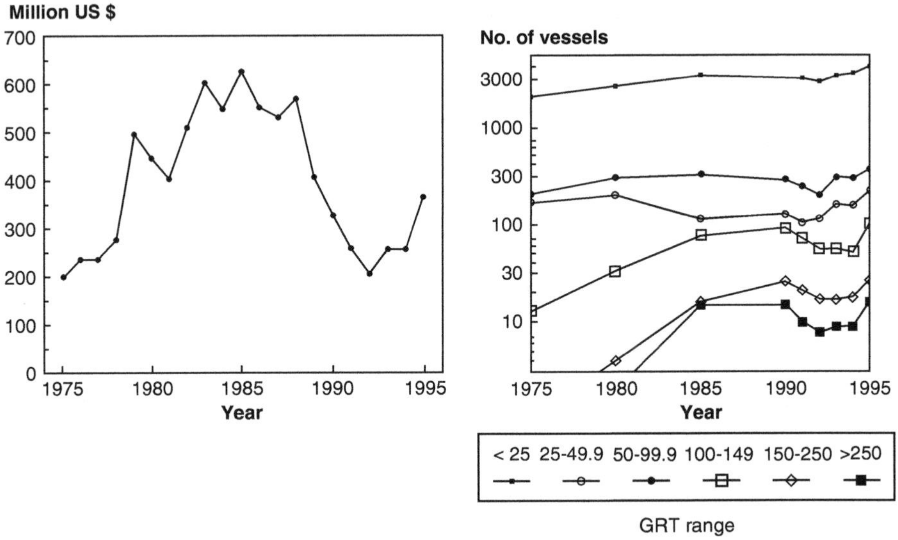

Changes in overall landed value and number of vessels operating in Black Sea fisheries, 1975–95 (UNEP sources) (figures from [18])

The technique in use to assess the economic cost during this period was to take an average pre-jelly figure of the annual value of the catches, such as $550 million, and subtract from it the post-jelly average of, say, $250 million. The resulting figure of $300 million a year represented the loss due to damage caused by the invading warty comb jelly. As catch information is not standardized and some catches are either valued in nonconvertible currencies or simply handled outside of the formal economy, these "lost catch" values differ to the tune of $200–400 million/year.

As the number of fishing vessels stayed more or less constant despite the much-reduced catches, some scholars began to add the value of the idled investment, at historic or new purchasing prices, or at depreciated value using different assumed annual amortization rates. This then becomes a lump sum, which is sometimes converted to an annual cost.

A third component that was sometimes included in the cost assessment is the opportunity cost of the foregone added value that processing industries would have given the landed fish, had they been caught in the first place.

Given these differences in what authors include in the costs and how they calculate them, it is no great surprise that the estimated losses range from $240 million to $1 billion annually [19–21].

Regardless of the figure used for the costs associated with the invasion of *Mnemiopsis leidyi* in the Black Sea, the real damage was of relatively short duration—a decade—since most of the effect, between 75 and 95% according to different estimates, ceased when *Beroe ovata* came on the scene.

Some Dissenting Views

The picture presented here gives the decisive role to the warty comb jelly in the collapse of Black Sea anchovy fisheries in 1989. This is surely the majority opinion, to judge from the literature, but there are reports suggesting that other, interacting factors might have been even more important or that the presumed cause-effect relationship was, in fact, the reverse.

Only because of severe overfishing and a remaining, much-reduced population of zooplankton-eating fish could *Mnemiopsis leidyi* mass develop by occupying a partly empty niche, goes the argument [7]. When fishing pressure eased in the early 1990s, the anchovy population partly recovered, and the jelly was pushed back, continues this reasoning.

Another cause argued to explain the large-scale changes in the Black Sea ecosystem in the late 1980s is climate change, in particular changes in atmospheric circulation patterns. Observed changes in the North Atlantic Oscillation in the last half of the 1980s were seen as a possible trigger for changes in plankton communities in seas in the Northern Hemisphere and thereby for changes in the population of plankton-eating fish. To test this, anchovy catches for the time period from the late 1980s to the early 1990s from many parts of the world were analyzed. It was found that the anchovy stocks from South Africa, the Benguela Current, California, and the Mediterranean had all collapsed at more or less the same time as the Black Sea anchovy stock had and that they had all recovered since mid-1992. The same paper also analyzed the effects of intensive fishing on the Black Sea anchovy stock.

The author of the study concluded that, "it is more likely that the collapse of the Black Sea anchovy, including the decrease of the zooplankton, was not a mere consequence of the *Mnemiopsis* invasion, but as well of overfishing and a shift in the oceanographic and meteorological regime at the end of the 1980s" [22].

With regard to assessments of the economic damage of *M. leidyi*, D. Knowler tried a much more strict, "theoretically defensible," in his own words, bio-economic analysis of the impact of the warty comb jelly, using a structural change approach. In his model, the optimum anchovy catch dropped from a pre-jelly figure of 400,000 t to a post-jelly one of 40,000 t and the profit of the fishing industry from $17 million a year to almost zero. Adding other species and "other valuable functions" of the fish species still leads him to the conclusion that "previous estimates of the losses in the anchovy fishery from *Mnemiopsis* may have been overestimated, perhaps by as much as 10-fold" [23].

Brief Comments on Some Other Areas with Invasive Warty Comb Jellies

In the spring of 2000, the warty comb jelly was first reported from the Caspian Sea. It was already present in several different locations, so it was obvious that the first arrival must have been at least a year of two earlier. Most probably, it came with ballast water in ships from the Black Sea. Mass development followed. In 2001, the densities already exceeded the highest ones ever reported from the Black Sea, and the following year they doubled again, with subsequent drastic effects on zooplankton and kilka/sprat catches [24]. These had declined from 400,000 t a year in the 1960s and 1970s to half of that in the 1990s, because of overfishing and pollution, and fell drastically again from 250,000 t in 1999 to an average of 67,200 t annually during the years 2001–2005, with a 54,300 low in 2005, because of *M. leidyi*. The decrease in the kilka stock was also affected by hydro-volcanic events, involving a significant gas blowout in the central and southern Caspian Sea in 2001 that caused mass mortality, estimated to have been 166,000 t of kilka [25].

Naturally, the successful control of the warty comb jelly by *B. ovata* in the Black Sea prompted speculations and investigations if it could do the same in the Caspian Sea, and in 2005 all five Caspian Sea states endorsed its introduction [26].

In 2006, the warty comb jelly was found in the North Sea and the western Baltic, and in the year after, it was also found in the central Baltic. It was first assumed that the route of invasion was boat transport from the Black Sea via the European canal system, but genetic evidence points to the source being from the northern distribution of this species in the western Atlantic (possibly Narragansett Bay) [27].

From the early 1990s, the occurrence of *M. leidyi* has been regularly reported in the Aegean Sea in the eastern Mediterranean, but it is unclear if the species became established there or if continuous influx from the Black Sea explained its presence. In 2009, blooms were reported from both Spain and Israel, indicating a Mediterranean-wide distribution [28]. Genetic studies pointed at multiple introductions, one undoubtedly from the Black Sea and another presumably from the Gulf of Mexico [27].

In the Mediterranean [29], North, and Baltic Seas [30], the *Beroe ovata* is present.

Summary

In the early 1980s, the warty comb jelly came to the Black Sea, probably with ballast water from the US East Coast. By the late 1980s, it had spread throughout the sea. In 1988, it was already common everywhere and had also entered the Sea of Azov, the most productive part of the water body. In 1989, the population exploded and reached a biomass that according to the most quoted source approached 1 billion tons wet weight, probably helped by the rich source of zooplankton that overfishing of anchovy had left for other zooplankton eaters to utilize. The mass development of the jelly led to the depletion of zooplankton, eggs and larvae of fish, mollusks, and crustaceans. A population crash of anchovies, horse mackerel, and sprat—the main zooplankton-eating fishes—was a consequence. A partial recovery of these and two further peaks of *Mnemiopsis leidyi* followed, with new ripples through the ecosystem. In 1997, another invasive jelly, *Beroe ovata*, a predator of the warty comb jelly, arrived. By 1999, this species had also spread throughout the Black Sea and mass developed, thereby almost eliminating its prey. Since then, the two species have shown a classic predator-prey balancing act, with only short periods of the year, with significant numbers of first *M. leidyi* and then *B. ovata*, and with limited impact on the zooplankton and fish populations.

Many figures circulate of the economic damage done by the invading warty comb jelly to fisheries in the Black Sea during its decade of dominance, and although the highest are the most quoted, they might not necessarily be the most reliable.

References

1. Kremer, P. (1976). Population dynamics and ecological energetics of a pulsed zooplankton predator, the ctenophore *Mnemiopsis leidyi*. In M. Wiley (Ed.), *Estuarine processes; Uses, stresses and adaptations to the estuary* (pp. 197–215). New York, NY: Academic Press.
2. Didžiulis, V. (2013). *NOBANIS—Invasive Alien Species Fact Sheet—Mnemiopsis leidyi*. From: Online Database of the European Network on Invasive Alien Species. Retrieved January 5, 2015, from NOBANIS www.nobanis.org
3. GESAMP (IMO/FAO/UNESCO-IOC/WMO/WHOI/IAEA/UN/UNEP Joint Group of Experts on the Scientific Aspects of Marine Pollution). (1997). *Opportunistic settlers and the problem of the ctenophore Mnemiopsis leidyi invasion in the Black Sea*. GESAMP Reports and Studies No. 58. London: International Maritime Organization.
4. Zaitsev, Y. P. (1992). Recent changes in the trophic structure of the Black Sea. *Fisheries & Oceanography, 1*(2), 180–189.
5. Borisov, V. I., & Kapitonov, E. I. (1973). *Sea of Azov* (in Russian). Baltimore, MD: KKI.
6. Kostianoy, A. G., & Kosarev, A. N. (2007). *The Black Sea environment*. New York, NY: Springer.
7. Eremeev, V. N., & Zuyev, G. V. (2007). Commercial fishery impact on the modern Black Sea ecosystem: A review. *Turkish Journal of Fisheries and Aquatic Sciences, 7*, 75–82.
8. Ivanov, L., & Beverton, R. G. H. (1985). *The fishery resource of the Mediterranean. Part II, The Black Sea, GFLM studies and reviews* (Vol. 60, p. 135). Rome: Food and Agriculture Organization of the United Nations.
9. Retrieved from http://www.encyclopediaofukraine.com/display.asp?linkpath=pages%5CA%5CZ%5CAzovSeaof.htm
10. Shiganova, T., Mirzoyan, Z., Studenikina, E., Volovik, S., Siokou-Frangou, I., Zervoudaki, S., Christou, E., Skirta, A., & Dumont, H. (2001). Population development of the invader ctenophore *Mnemiopsis leidyi*, in the Black Sea and in other seas of the Mediterranean basin. *Marine Biology, 139*(3), 431–445.
11. Konsulov, A., & Kamburska, L. (1998). *Ecological determination of the new Ctenophora - Beroe ovata invasion in the Black Sea* (pp. 195–197). Varna: Institute of Oceanology.
12. General Fisheries Council for the Mediterranean. (1993). Statistical bulletin, nominal catches 1979–1991. *FAO Bulletin Fisheries Statistics, 32*(9), 237.
13. Retrievedfromhttp://www.grida.no/graphicslib/detail/total-fish-catch-in-the-black-sea-over-the-last-30-years_06f0
14. Shiganova, T. A. (1998). Invasion of the Black Sea by the ctenophore *Mnemiopsis leidyi* and recent changes in pelagic community structure. *Fisheries Oceanography, 7*, 305–310.
15. Shiganova, T. A., Dumont, H. J., Mikaelyan, A., Glazov, D. M., Bulgakova, Y. V., Musaeva, E. I., Sorokin, P. Y., Pautova, L. A., Mirzoyan, Z. A., & Studenikina, E. I. (2004). Interactions between the invading ctenophores *Mnemiopsis leidyi* (A. Agassiz) and *Beroe ovata* Mayer 1912, and their influence on the Pelagic ecosystem of the Northeastern Black Sea Aquatic Invasions in the Black, Caspian, and Mediterranean Seas. *NATO Science Series, IV: Earth and Environmental Sciences, 35*, 33–70.

16. Kambuska, L., Moncheva, S., Konsulov, A., Krastov, A., & Prodanov, K. The invasion of *Beroe ovata* in the Black Sea a warning signal for ecosystem concern. Retrieved January 13, 2015, from http://www.academia.edu/3708091/The_Invasion_of_Beroe_ovata_in_the_Black_Sea_a_warning_signal_for_ecosystem_concern
17. Finenko, G. A., Romanova, Z. A., Abolmasova, G. I., Anninsky, B. E., Svetlichny, L. S., Hubareva, E. S., Bat, L., & Kideys, A. E. (2003). Population dynamics, ingestion, growth and reproduction rates of the invader Beroe ovata and its impact on plankton community in Sevastopol Bay, the Black Sea. *Journal of Plankton Research, 25*(5), 539–549.
18. Caddy, J. F. (2006). Recent experience and future options for fisheries assessment and management in the Black Sea: A GFCM perspective. (Informal document of the Black Sea Commission).
19. Caddy, J. F. (1992). *Rehabilitation of natural resources. Environmental management and protection of the Black Sea.* Technical Experts Meeting, 20–21 May, Constanta, Romania.
20. Campbell, D. (1993). *Socio-economic study of the Black Sea fisheries.* Report of the Second Technical Consultation on Stock Assessment in the Black Sea, Ankara, Turkey, 15–19 February. FAO Fisheries Report No. 495, General Fisheries Council for the Mediterranean, FAO, Rome.
21. Travis, J. (1993). Invader threatens Black, Azov Seas. *Science, 262*, 1366–1367.
22. Niermann, U. (2004). Mnemiopsis leidyi: Distribution and effect on the Black Sea ecosystem during the first years of invasion in comparison with other gelatinous blooms. *NATO Science Series, IV: Earth and Environmental Sciences, 35*, 3–31.
23. Knowler, D. (2005). Reassessing the costs of biological invasion: Mnemiopsis leidyi in the Black sea. *Ecological Economics, 52*, 187–199.
24. Shiganova, T. A., Dumont, H. J., Sokolsky, A. F., Kamakin, A. M., Tinenkova, D., & Kurasheva, E. K. (2004). Population dynamics of Mnemiopsis leidyi in the Caspian Sea, and effects on the Caspian Ecosystem. *NATO Science Series, IV: Earth and Environmental Sciences, 35*, 71–111.
25. Daskalov, G. M., & Mamedov, E. V. (2007). Integrated fisheries assessment and possible causes for collapse of anchovy kilka in the Caspian Sea. *ICES Journal of Marine Science, 64*, 503–511.
26. Retrieved from http://www.payvand.com/news/05/may/1162.html 21/5 2005
27. Ghabooli, S., Shiganova, T. A., Briski, E., Piraino, S., Fuentas, V., Thibault-Botha, D., Angel, D. L., Cristescu, M., & MacIsaac, H. J. (2013). Invasion pathway of the Ctenophore Mnemiopsis leidyi in the Mediterranean Sea. *PLoS One*. Retrieved January 16, 2015, from http://journals.plos.org/plosone/article?id=10.1371/journal.pone.0081067
28. Fuentes, V. L., Angel, D. L., Bayha, K. M., Atienza, D., Edelist, D., Bordehore, C., Gili, J.-M., & Purcell, J. E. (2010). Blooms of the invasive ctenophore, *Mnemiopsis leidyi*, span the Mediterranean Sea in 2009. *Hydrobiologia, 645*(1), 23–37.
29. Galil, B. S., Gevili, R., & Shiganova, T. (2011). Not far behind: First record of Beroe ovata Mayer, 1912 (Ctenophora: Beroida: Beroidae) off the Mediterranean coast of Israel. *Aquatic Invasions, 6*(Suppl 1), S89.
30. Shiganova, T. A., Riisgård, H. U., Ghabooli, S., & Tendal, O. S. (2014). First report on Beroe ovata in an unusual mixture of ctenophores. *Aquatic Invasions, 9*(1), 111–116.

Some Concluding Remarks Emanating from the Case Studies

Aliens forever or integrated immigrants with time?
or

Will species introduced by man ever be seen as native?

Rachel Carson, as mentioned in the chapter on "Starlings in North America," argued that that bird ought to be given "citizenship papers" in the USA, in recognition that it can by now be considered "native." That hasn't happened yet, but it might someday, if the starlings hang on. Many other species introduced by man have, with time, become seen as being an integral part of local nature. Among the species with chapters in this book, earthworms in North America are perhaps closest to having achieved this status (most people are completely unaware that they were introduced from Europe centuries ago). Other examples that have been mentioned, albeit more in passing, are dingoes (that probably came to Australia from Indonesia with seafarers some 4000 years ago) and rabbits (brought to the British Isles by the Normans in the twelfth century). Both are today seen as native, although the latter is estimated to cause agricultural damage in the UK to a value of $ 150 million annually. Other well-established immigrants in the UK that by now seem to have received their citizenship papers are the horse chestnut tree (*Aesculus hippocastanum*), which arrived in the sixteenth century, and Reeves' muntjac (*Muntiacus reevesi*), commonly called "barking deer," a relative newcomer with less than a century on the isles.

In Sweden, on the Scandinavian Peninsula, the European hare (*Lepus europaeus*) was first introduced by Queen Kristina to an area outside of Stockholm, around 1650. Further introductions followed during subsequent centuries with especially large numbers in 1886–1887 in the southern Swedish province of Skåne. The European hare has now colonized most of Southern Sweden and Norway. Despite having had a role in the decline of the native mountain hare (*Lepus timidus*), it has now more or less gained acceptance, as indicated in the shift of its popular name from tyskhare (German hare) to fälthare (field hare). Other accepted newcomers are fallow deer (*Dama dama*) and pheasant (*Phasianus colchicus*), which arrived in the 1570s and eighteenth century, respectively, followed by many more introductions to various places in Southern Sweden, before the populations became firmly established.

A. Jernelöv, *The Long-Term Fate of Invasive Species*,
DOI 10.1007/978-3-319-55396-2_19

Another "accepted" invader in most of Northern Europe is the lupine (*Lupinus polyphyllus*), from Western North America. It was imported as a horticultural plant and escaped the confines of the gardens in the nineteenth and early twentieth century.

A much-sung native plant (e.g., by the poet William Wordsworth) in England, and a somewhat threatened and protected one in Germany, the wild daffodil (*Narcissus pseudonarcissus*) is a beloved import to Scandinavia and is common in the southern parts of Sweden, Norway, and Finland. It has been present in the Fennoscandian region at least since the botanist and zoologist Carl von Linné first saw it and named it in 1768.

Yet another "newcomer" that has received citizen status is the Roman snail (*Helix pomatia*), which the Romans brought to the British Isles almost 2000 years ago. In England, the Roman snail is a protected species under the Wildlife and Countryside Act of 1981. To Sweden, monks and French-minded aristocrats imported them from the Middle Ages to the eighteenth century, and it now has a treasured, somewhat spotted, distribution in the south of the country. It can generally be found in the vicinity of old monasteries and castles or country mansions, to where it was brought, and it has survived over the centuries, but not spread much on its own accord.

I have my own personal story of the further distribution of the Roman snail, which in Sweden is called vinbergssnäcka or "wine mountain snail." As a 10-year-old boy, I lived in a suburb south of Stockholm in a house on land that had once been part of the grounds of a mansion called Långbro Gård, which had belonged to the Oxenstierna family in the seventeenth century. There were numerous wine mountain snails in the vicinity, and I collected them and fought for their protection against other neighborhood boys, who liked to throw them against rocks and smash them. When I turned 11, my parents bought a summer house in the Stockholm archipelago, in a sparsely settled, mostly wooded area called Söderby, along the shore of the Baltic at the Harsfjärden Fjord. It was clear that I would be away for the whole of the school summer vacation and, fearing my absence would leave my beloved snails defenseless, I decided to rescue them. To this end, I filled two buckets with wine mountain snails and brought them to the summer house, where I released them. For the first week or so thereafter, I could still find some now and then, but then they disappeared. I thought my rescue project had failed. Five years later, though, a summer house neighbor reported in surprise, not knowing about my earlier actions, that he had accidentally stepped on one. After 10 years, wine mountain snails were being spotted regularly. Twenty years later, they were ubiquitous. Sixty years later, with much of the area having been developed in the interim, they are again sparse, but can always be found among the long grasses and wildflowers in road verges on a wet summer day. On a small local scale—they don't seem to have spread more than half a kilometer from the point of release in the early 1950s—they have behaved just as several of the truly invasive species in the chapters of this book. (By the way, they are now gone from the Långbro Gård area where I had found them as a child.)

Along the Atlantic coast of Europe, the harvests of the European oyster (*Ostrea edulis*) had dwindled due to overfishing and disease in the 1950s and 1960s. From 1970 to 1973, French oyster farmers brought in a massive amount of "spat" (baby Pacific oysters) from Japan and hundreds of tons of mature Pacific oysters (*Crassostrea gigas*) from British Columbia [1]. *C. gigas* saved the European oyster industry, but it also spread well beyond the confines of the aquaculture areas into which it had been introduced and became an invasive species. One of the real worries in Southwestern Europe is that it could wipe out the Portuguese oyster (*Crassostrea angulata*), which has been seen as a regional treasure on the Iberian Peninsula. However, the Portuguese oyster is also an import from the Pacific, and many scientists see it as a subspecies of *C. gigas*—the two easily interbreed. In the 400 years or so since Portuguese seafarers and ships brought it to European waters, the Portuguese oyster has obviously become an accepted immigrant.

While the attitude to newcomers is becoming progressively more negative, the urge to reintroduce species that died out in a country has become stronger and stronger. I was myself responsible for bringing the white stork (*Ciconia ciconia*) back to Sweden in 1979, when it had been gone as a nesting bird since 1954. Mostly, however, the reintroductions are of species with a far longer absence. The European beaver (*Castor fiber*) was hunted to extinction in Sweden in 1871 and reintroduced 50 years later with specimens from Norway. The current population is estimated to be 100,000 individuals and growing.

Beavers have had a much longer wait for return on the British Isles, from which they had been gone for some 400 years when reintroduction started in 2009.

The wild boar also had to wait a while to come back to Sweden, and its reintroduction had political implications. It became extinct in the early eighteenth century through intensive hunting. Sweden's King Fredrik I first reintroduced them to the Baltic island of Öland in 1723, over the vociferous protests of local farmers. Immediately after the king's death, the Swedish parliament took the decision to re-exterminate the wild boars, and the decision was implemented in 1752. A century passed before, as a private initiative in the 1870s, a man named Oscar Dickson had wild boars brought to a hunting enclosure on his mansion at Skeppsta, south of Stockholm. Later, several other aristocrats followed suit. Escapees from such enclosures gave rise to a few small Swedish wild boar populations. The Swedish parliament in 1980 reiterated the decision from 1752 that the wild boar was unwanted in Sweden and imposed on the authorities to rid the country of them but reversed that decision in 1987, declaring that the wild boar was a "natural member of the Swedish fauna." Since it got back its citizenship papers, the wild boar population has expanded to some 300,000 individuals and continues to grow. It is now present in most of Southern Sweden—to the dismay of farmers.

One of the most spectacular reintroductions ever has been that of the European bison (*Bison bonasus*), also known as the wisent. The species, extinct in the wild with only some 50 animals alive in zoos, a third of which at Skansen in Stockholm, was successfully reintroduced with a start in the Białowieża Forest in Poland in 1951. Currently, the total number of animals exceeds 6000, and free-ranging herds are present in ten countries. In some of those, the wisent had been gone for more than 500 years [2–4].

When a species has been gone for a long time, the ecosystem to which it once belonged will have undergone change. From a human individual's point of view, the thus modified system may be regarded as the natural one, the one that existed "in all of human memory." The changes and adjustments that will follow a successful reintroduction may thus be perceived in a similar way to those following an alien invasion. Beavers, wolves, and wild boars bear witness to this. The European bison—the wisent—may soon share the experience.

In most countries, there is no official policy with regard to the length of time a species has to be present before it more or less automatically receives citizenship or is considered native. Where policies exist, they can differ significantly. As noted in the chapter on the "Himalayan balsam," in the Czech Republic, plants introduced after anno 1500 are termed "neophytes," denying them the real right to belong. In the chapter on the "Raccoon in Europe," a section of German legislation is referred to in which it is stated that an animal or plant that "in nature, without human support, can survive in several generations, is to be regarded as native ('heimisch'), also when it was first introduced by Man." That, of course, can happen within a decade for many species.

The long-term trend in attitude toward new species in a territory ranges from "enrichment societies" and individuals with solid scientific credentials trying their best to acclimatize desired species from the Old World to the New or vice versa to national policy decisions and international agreements aimed at keeping all aliens out and eradicating those that have gotten in. A prime example is the list of 37 species that the European Union is poised to pass as a follow-up and concretization of a "principle decision" from 2014 [5], meaning it will carry the provision that EU member states must eradicate them, if at all possible. In that, the EU administrators and politicians are in line with the opinion of most scientists in the field. The scientific community is, however, not unified on the subject, as illustrated by a heated debate in 2011, in the scientific journal *Nature*. In an article by Davies et al. [6] headed "Don't Judge the Species on Their Origins," the authors argued that there is nothing that indicates that the spread of species will slow down, but there are many signs that it will accelerate as we move further into the Anthropocene epoch. The same article held that nothing that points to future eradication campaigns will be any more successful or less costly than attempts made in the past and that we therefore should focus on controlling the relatively few invasive species that cause real problems. A reply by Simberloff and a host of others [7], published in the following issue, was titled "Non-natives: 141 Scientists Object," rejecting the notion and essentially arguing for border controls and early eradication before aliens can get a foothold. As a member of the editorial group of an environmental science journal, I now and then see articles similar to Davies' being submitted—and mostly rejected—after critical reviews from Simberloff's sympathizers.

Quite often, as pointed out in a paper by Larson [8], the language in articles on nonnative species, even in the scientific press, reveals a bias on behalf of the authors that run counter to the traditional ideal of objectivity and neutrality on part of natural scientists. The aliens are often called "enemies," "intruders," or "rapists" (if they cross-fertilize), the front of spread is named the "battle zone," and nonnative plants "bombard" the natives with their seeds.

As mentioned in some chapters of the book, environmental NGOs often engage in control and extermination campaigns, thereby aligning themselves with the larger scientist group. The general public, however, often has a more positive view of the aliens, and newspapers and journals often support them, as did *The Economist* in two articles in 2015 with the titles "In Praise of Invasive Species" and "Invasive Species—Day of the Triffids" [9, 10]. In many ways, it feels like a repetition of the American sparrow war of the 1880s.

Delayed Invasiveness

The 18 case studies in this book have one feature in common: the population explosions took place a short time after introduction. This, no doubt, is the most common situation, but there are also a number of cases where a longer period of adaptation was required before the alien species acquired the competitiveness that made them invasive. Such cases offer easy-to-see examples of how alien species adapt and evolve in their new environments. In most such cases, it is domesticated plants and animals that first become barely surviving alien garden escapees or feral populations that have first to regain some traits of their wild ancestors before they become invasive. Feral dogs, cats, and hogs are examples of such animals, whereas Johnson grass (*Sorghum halepense*), air potato (*Dioscorea bulbifera*), and Hottentot-fig (*Carpobrotus edulis*) provide similar examples of plants.

Of the case studies in this book, the one on the "European rabbit in Australia" touches on this subject when noting that domesticated rabbits for farming were introduced to Australia well before the wild ones came as game animals. It is generally assumed that the latter formed the basis of the so-called rabbit "invasion," even as escaped domesticated rabbits might have contributed to an increased genetic diversity and better further adaptability of the invading population. The Tasmanian rabbit population from the late 1820s, however, might be of domesticated origin.

Another brief touch on the concept of delayed invasiveness—a reverse one—is to be found in the chapter of the "American mink in Europe." There, it is noted that the original caged minks, with many still retaining their wild characteristics, quickly formed invasive populations when they escaped, whereas modern farmed minks, in addition to being much larger, have been domesticated to the degree that they seem to have lost a good deal of their invasive capacity. The fact that this has taken only about a century is noteworthy.

Another reason for delayed invasiveness is that an alien species that initially meets an environment where it can merely hang on, later finds that conditions from its point of view improve and render the species competitive. A case in point is that of escaped garden plants in Northern Europe. These often originated in warmer areas, so milder winters in particular, such as the region has experienced during the last decades, make them more competitive. Snowdrops may be one such example, as alluded to in the chapter on "Himalayan balsam." Likewise, the current extended dry spell in California has added aggressiveness to alien desert plants such as the Sahara mustard (*Brassica tournefortii*).

Ecological Effects of Invasive Species

Is there any general conclusion that can be drawn from the 18 case studies with regard to the ecological effects of the invaders? In reality only one, as I see it: once in a new environment, the invasive species will start to adapt. In, say, ten generations, it will no longer be quite the same. Resident species will change, and to some extent the chemical and physical properties of the environment will also alter, so a new, successful immigrant will bring about a modified ecosystem. In noting that, it is important to remember that an ecosystem is never a fixed structure but an undulating and evolving entity. However, an invasive species is likely to cause larger waves than the usual ones, at least to start with.

Another reoccurring phenomena, excluding among the case studies the two pathogens and the Colorado potato beetle, but including the extremely successful earth worms in North America, are the initial overshoot—the peak in the numbers and density of the invasive species a relative short time after introduction—and the subsequent decline or normalization. In that there is an interesting resemblance, I find, with the processes of recovery of heavily polluted bodies of water. Massive oil spills provide examples of this as do many industrially polluted waters after cleanup.

One example from my experience is the small lake Bålsjön in Bergslagen in middle Sweden. An iron ore dressing plant operating on its shore closed down in the late 1960s. Despite the fact that iron in the ore was present in two forms, magnetite and hematite, the old plant separated out only the magnetic part, discharging the other together with the crushed and pulverized rock. While settling ponds would capture most of the latter, the hematite ended up in the lake, giving the water the color of rosehip soup and a Secchi disk transparency of only a few centimeters. Naturally, the lake was almost devout of higher life forms, with the noticeable exception of very small, but sexually mature noble crayfish, called "bumblebees" by the locals, who spotted them when they at dusk crawled up on land to feed on horsetail and coltsfoot that grow richly on the sides.

The plant closed in autumn and throughout the winter not much seemed to happen under the ice, except that a layer of clear surface water under the ice slowly became deeper as the hematite particles settled. In spring, when the snow on the ice melted, we found that on the more and more rugged underside of the ice, a thick layer of attached diatoms developed. When the ice melted in May, the algae became planktonic and an early bloom was kick-started. A spring bloom of diatoms is not at all unusual in Swedish lakes. Normally they are followed by green algae as dominant phytoplankton in early summer and

blue-greens in late summer/early fall, not so in Bålsjön in 1969. The diatom bloom continued throughout the year giving the lake water a dingy color and a low transparency.

In 1970, the story was initially repeated but in June came a great surprise. An ostracod, a very small crustacean with the Latin name *Notodromas monacha*, that normally in small numbers live on or at waterplants—of which there were hardly any in Bålsjön proper—and not free-swimming in the open water, mass-developed as a zooplankton. In the literal zone, from the beach and five meters out, the densities were staggering. Around midsummer time, we estimated the ostracod biomass in the lake to some 30 tons. However, the mass development of the small crustacean didn't last for long. By September it was over. By then small bleaks, up to 7 cm in length, dominated the lake fauna. For one-summer bleaks in Sweden, they were unusually large, and with their high numbers and large shoals, they reminded mostly of herrings in their behavior.

The high growth rate the bleaks had had the first summer was replaced by hardly any growth the following year. In the spring, when we found lots of strings of perch roe, we expected that the dominance of the small bleaks would be of short duration and that they soon would fall prey to a large year class of perch, one of their natural predators. That, however, didn't happen. The one year older bleaks practically eat the lake clean of perch fries. They ate their own offspring also and the 1970 year class of bleaches, never growing bigger than 10 cm, dominated the fish fauna for 5 years. By that time, some very large perches and pikes had been caught in the lake together with a burbot (*Lota lota*). The scales we could analyze from some of them revealed that they had had a normal growth rate for their species in the general area for the first 3–4 years of their lives and that it after that exploded. They obviously had found the way to Bålsjön as adults and been able to feed well on the bleaks and grow big. As long as the bleaks remained so numerous, however, and with little plant cover in addition, hardly any of their offspring survived.

None of the species involved in this turbulent recovery of the heavily polluted lake was alien, and none was brought there by man, but some patterns in the ecological process resemble those that invasive species can cause.

In this book, the invasion into new lands of 18 different species is described, ranging from plants to animals, from aquatic to terrestrial, from grazers to predators, and from invertebrates to mammalians and birds, including also two pathogenic fungi. It is self-evident that their specific ecological effects are very different. Despite that, however, with the changes in the recipient ecosystem, the mass development of any new species could be expected to be at the expense of (some of) the old ones. How this "at the expense of" plays out may differ. A classical example can be found in Northern Scandinavia, in mountain lakes with only one species of fish naturally present—the Arctic char (*Salvelinus alpinus*). It thus occupies the entire "fish niche," feeding on insects on the waters' surface and on bottom fauna, as well as on everything in between. Initially the Sami people, and later also sport fishermen, introduced trout (*Salmo trutta*) into such lakes. One result, where the trout has survived, is that the two species divide the lake into two "fish niches" with respect to food, with the trout largely feeding on the surface and the char on the bottom. Another is that the char biomass decreases, but the fish biomass goes up. Traditional Chinese pond aquaculture with multiple fish species is based on this concept.

It is thus not inevitable that new species replace the existing ones, but such thinking seems to be what has led to the routine, some might say "knee-jerk" statement on many so-called "fact sheets" on invasive species, that the presence of an invasive species invariably leads to "extinction of native species and loss of biological diversity."

With regard to the claim of loss of native species, none of the 18 case studies has the extinction of any native species been directly coupled to an invader. Reduced abundance and (temporary) local absence have been regular features, but not outright extinction.

The question of loss of biological diversity is more difficult to answer because the term itself is quite complex and multifaceted. The Convention on Biological Diversity (CBD) uses the following description: "The variability among living organisms from all sources including, inter alia, terrestrial, marine and other aquatic ecosystems and the ecological complexes of which they are part, this includes diversity within species, between species and of ecosystems."

Leaving aside, in this context, genetic diversity within species and ecosystem diversity, and focusing on species diversity, there are two main components of the concept: number of species and number of individuals. There are some 30 different formulas in use to calculate species diversity, and most of them are associated with slightly different definitions. What they all have in common is that the number of species affects the diversity positively and so does an even distribution of the number of individuals over the species. Examples are Simpson's and Shannon's indexes:

In Simpson's diversity index (D), the proportion of species (i) relative to the total number of species (p_i) is calculated and squared. The squared proportions for all the species are summed, and the reciprocal is taken:

$$D = \frac{1}{\sum_{i=1}^{S} p_i^2}$$

In Shannon's diversity index (H), the proportion of species (i) relative to the total number of species (p_i) is calculated and then multiplied by the natural logarithm of this proportion ($\ln p_i$). The resulting product is summed across species and multiplied by -1:

$$H = -\sum_{i=1}^{S} p_i \ln p_i$$

There are very few cases where any species diversity index has been calculated for an area with and without an invasive species. Rough attempts to estimate the direction of change in species diversity indices, using only available data on species richness and impressions of evenness, do not yield any results showing a general decrease following invasions. (An unanswered, principal question is whether the invader itself and other aliens shall be included in the index. Doing so mostly increases species richness, while decreasing evenness. Excluding all nonnatives generally has the opposite effects.)

Thus, the routine statement of loss of biological diversity following invasions seems not to be well supported by data.

Assessments of Economic Effects of Invasive Species

Three striking observations can be made immediately, when one goes through a number of studies of the economic effects of invasive species:

1. Damage attracts much more attention than benefits.
2. Different estimates of the economic effects of an invasive species in a new area spread over a wide range—seldom less than an order of magnitude (a factor of 10).
3. The high damage estimates are quoted much more often than the low ones.

One question that authors handle in very different ways is that of reduced turnover or reduced profit. There are many examples of this both with regard to fisheries, agriculture, and forestry. Suppose that an alien grazer damages one type of crop in a given year to the extent that in the following years, if the farmers see that the invaders are still numerous, they don't plant that crop. Is the damage calculated as the value of the nonproduced crop or the profit (sales value minus expenses for things like fertilizers, diesel oil, salaries, etc.) that is foregone? Depending on location and crop, such figures may well differ with a factor of 10.

Another question relates to volume and value. If, e.g., an introduced aquatic species reduces fish catches from 100,000 tons to 30,000 tons, but the price of the fish, because of the scarcity, triples from $2000 to $6000 per ton, what's then the economic damage? Some authors count it as $ 20 million (100,000 tons × $2000–30,000 tons × 6000), others $140 million (70,000 tons × $2000) or even $420 million (70,000 tons × $6000). Add to that the question, if the fishermen's expenses should be deducted or not, (same question as in the agricultural example above) and the different methods for damage estimate will give results that easily span two orders of magnitude (a factor of 100).

A third question relates to what percentage of the individuals of a species that can be assigned the same value as that of a few, e.g., is what you pay for a few barnacles in a zoo shop a relevant per capita value to be applied to the millions on a rocky beach? Or is the value of a stately ornamental tree in a park relevant for all trees of the same species and size, regardless of where they are to be found?

Calculations of economic benefits are likewise beset with more or less arbitrary methods. Is the value of commercial compost soil multiplied with the amount of similar soil that earthworms produce in North American agricultural areas a good indicator of the economic benefits that earthworms bring to North American agriculture?

Is the relationship in the USA between the value of the honey that bees produce and the value of the pollination services they provide for commercial crops also relevant for Latin American countries?

A question relevant both on the negative and positive sides of the scale is if changes in property values, as a consequence of environmental changes brought about by an invasive species, should be taken into account and if so, how.

An obvious conclusion of what is said above is that estimates of the economic effects of invasive species should be taken with several grains of salt and that strict methods and agreed definitions are badly needed. Efforts in this direction have been made some of which are highlighted in the species chapters, but they are still very far from being in general use.

What Can We Expect from Ongoing Invasions?

Is there anything we can learn from the 18 case studies in this book that can shed some light on the long-term fate of some of today's most troublesome invaders?

Of the 18, one can be said to truly have conquered the new land: earthworms in North America. In the process, they have changed the composition of the continent's soils, literally preparing the ground for other Euro-Asian species including agricultural crops, and have thereby brought major economic benefits.

Another very successful invader, to be found at the opposite end of the economic scale thanks to man's cultivation of potatoes, is the Colorado potato beetle that is still spreading and causing very significant economic damage. The same goes for one of the fungi causing Dutch elm disease, but not for the other, which, within 30 years, lost most of its potency.

Another species, also very successful but on a smaller scale, and with the immediate threat of a human counter-offensive hanging over them, is the beavers on Tierra del Fuego. Also they have transformed the landscape, but with the negative economic impacts arguably at least matched by the positive ones.

Holding on to conquered territory, but apparently not expanding it much anymore, is the Argentine ant on Mediterranean shores.

Other invaders with long-term success and continued expansion are the Himalayan balsam in Europe and, on a smaller geographic scale, the raccoon.

An invader that is almost holding its own despite governmental control attempts, after having spread over much of North America, is the starling.

A somewhat special case is the African "killer bee" in the Americas, which can be said either to have drowned in the gene pool of the previously introduced European honey bee or to have conquered the continent by injecting its genetic traits into the same.

In three cases, the rabbits in Australia, the water hyacinths in Africa and Asia, and the warty comb jelly in the Black Sea have been brought under control through the introduction of pathogens, grazers, and predators.

Of the remaining species, the zebra mussel and the marine toad are still expanding their geographical range within the new territories but are simultaneously experiencing substantial population declines in areas they conquered earlier.

Dwindling numbers are also the trend for initially successful invaders such as the Canadian waterweed, the house sparrow, and the American mink.

Obviously, the outcomes so far span a wide range with few common features. One of those, though, common not for all but for most of the species in this study, is that the initial population explosion that brought the invader at its peak to huge numbers and high population densities is followed by a more or less rapid decline. Another feature of the successful invaders is their staying power. Even if their numbers go down to a trickle of what they were at their peak, these species do not die out. The Canadian waterweed in Europe is a telling example of this.

The lack of native predators is often seen as a reason why alien species become invasive. Mostly though, some members of the native fauna adapt to feed on the newcomers. Birds and moles eat earthworms in North America; eagles and dingoes hunt rabbits in Australia; crabs and roaches feast on zebra mussels, and deer and aphids consume the Himalayan balsam. Such consumption may not be enough to control the species within the time span of human patience, but development of local predators is a factor that becomes more and more important with time. In addition, as noted in the summary above and in several of the chapters, introduced alien predators or pathogens can often do the job.

Another, at least as important factor behind the invasive success of aliens, to judge from these case studies, is the lack of competition. The American minks would probably not have found Europe as welcoming had foxes and otters, not to talk about the European minks, been there in numbers. Likewise, the warty comb jelly might have had a tougher time conquering the Black Sea had overfishing not reduced the local plankton-eating fish so much. Massive recovery of such suppressed native species might help bring the invader under control.

The main lesson from these case studies, however, is that there are more unique features than common ones. This makes it difficult, bordering on impossible, to foresee the long-term fate of most current invaders with any precision. Thus, Japanese knotweed (*Fallopia japonica*), silver carp (*Hypophthalmichthys molitrix*), or yellow starthistle (*Centaurea solstitialis*) may still be seen as troublesome invaders in North America a century on, or they might have a low-key, hardly noticeable, presence.

In some cases one might dare a guess based on the case studies. The golden mussel (*Limnoperna fortunei*) that came from Asia to Latin America and is now spreading into the river systems of Argentina and Brazil has so many features in common with the zebra mussel that it can be seen as a tropical equivalent. Thus, one would expect an intensive invasion causing significant ecological changes and economic damage, but one that would peak within a decade of establishment in any specific area.

Even more of a guess concerns the "murder snail," as the Spanish slug (*Arion vulgaris*) is called in Scandinavia, where it invades gardens and damages plants. It is closely related to and interbreeds with the native black slug (*Arion ater*), (which means that when a species was defined as a "closed gene pool," the two would not have counted as different species). There are not many traits that the Spanish slug shares with the African killer bee, but they are both southern (sub)species brought into contact with northern potential sex partners. Based on this alone, one might guess that the Spanish slug will drown in the gene pool of the black slug, or if one prefers an alternative way of saying the same thing, it will take over by having some of its genes permanently established in the entire forest slug population.

A Bleak Future (?)

One thing, though, can be said with a good deal of certainty: The flow of alien species to practically every part of the world will continue—probably at an accelerating pace—following growth in international trade and travel. To prevent invasions of these aliens, import controls and early extermination programs would have to be developed and installed at an unprecedented—and almost unimaginable—level that would certainly not come cheaply. Whether this is even possible, let alone probable, is at the heart of the 2011 Nature debate. Should such serious efforts not come about or if they should fail, it is possible that we have only seen the beginning of a mega-trend: The breakdown of barriers to free mobility of all species, where the question will not be "can they get here" but "can they survive here." Ecosystems in the same climatic zone all over the world might then come to share most of the species and be much more similar to each other than they are today.

If so, the German definition of "heimisch" (native) as a species that has survived in the country in nature without human support for several generations may become the international standard. Control programs would then have to focus on staving off population explosions of the real villains—however they are defined and whatever claim they have to citizenship.

References

1. McCabe, J. *Oysters in the 20th century, France*. Retrieved from http://www.oysters.us/france-20.html
2. *How the near-extinct European bison made a comeback*. Retrieved 10 April, 2010, from TheDodo.com
3. *Wisents in Slovakia: The population has increased three times since 2004*. Retrieved July 23, 2013, from European Wildlife
4. *Die Wisente kehren nach Deutschland zurück—Wissen & Umwelt*. Deutsche Welle. 11 April 2013.
5. Retrieved from http://eur-lex.europa.eu/legal-content/EN/TXT/?uri=OJ:JOL_2014_317_R_0003
6. Davis, M., et al. (2011). Don't judge species on their origins. *Nature, 474*, 153–154.
7. Simberloff, D., et al. (2011). Non-natives: 141 scientists object. *Nature, 475*, 36.
8. Larson, B. M. H. (2005). The war of the roses: Demilitarizing invasion biology. *Frontier in Ecology and the Environment, 3*, 495–500.
9. (2015, December 5–11). In praise of invasive species. *The Economist* (p. 18).
10. (2015, December 5–11). Invasive species—Day of the triffids. *The Economist* (pp. 53–54).

Index

A. Jernelöv, *The Long-Term Fate of Invasive Species*,
DOI 10.1007/978-3-319-55396-2

MIX
Papier aus verantwortungsvollen Quellen
Paper from responsible sources
FSC® C105338

If you have any concerns about our products,
you can contact us on
ProductSafety@springernature.com

In case Publisher is established outside the EU,
the EU authorized representative is:
Springer Nature Customer Service Center GmbH
Europaplatz 3, 69115 Heidelberg, Germany

Printed by Libri Plureos GmbH
in Hamburg, Germany